CHRONOTHERAPY

Michael Terman, Ph.D.

Director, Center for Light Treatment
and Biological Rhythms
Columbia University Medical Center

Ian McMahan, Ph.D.

City University of New York

AVERY A MEMBER OF PENGUIN GROUP (USA) INC. NEW YORK

CHRONOTHERAPY

Resetting Your Inner Clock to Boost Mood, Alertness, and Quality Sleep

Published by the Penguin Group
Penguin Group (USA) Inc., 375 Hudson Street, New York, New York 10014, USA •
Penguin Group (Canada), 90 Eglinton Avenue East, Suite 700, Toronto, Ontario M4P 2Y3, Canada (a division of Pearson Penguin Canada Inc.) • Penguin Books Ltd, 80 Strand, London WC2R 0RL, England • Penguin Ireland, 25 St Stephen's Green, Dublin 2, Ireland (a division of Penguin Books Ltd) • Penguin Group (Australia), 707 Collins Street, Melbourne, Victoria 3008, Australia (a division of Pearson Australia Group Pty Ltd) • Penguin Books India Pvt Ltd, 11 Community Centre, Panchsheel Park, New Delhi–110 017, India • Penguin Group (NZ), 67 Apollo Drive, Rosedale, Auckland 0632, New Zealand (a division of Pearson New Zealand Ltd) • Penguin Books, Rosebank Office Park, 181 Jan Smuts Avenue, Parktown North 2193, South Africa • Penguin China, B7 Jaiming Center, 27 East Third Ring Road North, Chaoyang District, Beijing 100020, China Penguin Books Ltd,

Registered Offices: 80 Strand, London WC2R 0RL, England

Published simultaneously in Canada

ISBN 978-1-58333-472-0

Printed in the United States of America

BOOK DESIGN BY ELLEN CIPRIANO

ALWAYS LEARNING PEARSON

What our book does and doesn't offer

Our goal in writing this book is to help readers make informed decisions about their health and the health of their loved ones. It should not be seen as a substitute for treatment by, or the advice and care of, a professional health care provider. We the authors and our publisher have done our best to make sure that the information we give you here is accurate and up-to-date. We are not engaged in rendering professional advice or services to the individual reader, and shall not be liable or responsible for any adverse effects or consequences that might be seen as arising from information or suggestions herein.

To JANE, JIUAN, and SELENA

Contents

PART 5. CHRONOTHERAPY IN YOUR LIFE

Foreword

John F. Gottlieb, M.D.

Feinberg School of Medicine
Northwestern University

Our current understanding of mental health is being challenged and enlarged by groundbreaking approaches. One of the most promising is psychiatric chronotherapy. This book offers both a thorough exposition and a new synthesis of this advance. But to understand how it emerged and why it is so important, we need some background.

Modern psychiatric research, training, and clinical practice are heavily influenced by the pharmaceutical industry. Complaints about the effect of Big Pharma can be found from within and outside the medical establishment. From the subtle effects of publication biases and journal sponsorship to the more overt pressures found in research funding directives, medication-based treatment algorithms, and industry-supported training, the hand of Pharma casts a broad shadow over the current practice of psychiatry. To be fair, the interests of this economic sector have also dramatically moved our field forward. New medications have been developed, treatments have become more effective, and our understanding of the neurobiology of mental illness has exponentially increased. Despite, or perhaps even because of, these advances, our field continues to lean toward neurotransmitter and receptor theories of dysfunction that can be corrected with new psychotropic agents. We live in an age of pharmacotherapy.

Against this background, the new field of chronotherapy is shouldering into view, offering a different set of ideas about what goes wrong in emotional and behavioral problems and what can be done to address it. Chronotherapy, in turn, owes its existence to the basic science of chronobiology—the study of circadian rhythms. Chronobiology is itself a relative newcomer. As recently as fifty years ago, circadian biologists hotly debated whether internally generated rhythms even existed in humans. It was thought that we had evolved past these ancient timing structures and that the ebb and flow of our cognitive, behavioral, and emotional processes was all socially determined. Through painstaking research and novel experimental methods, humans were found to share a circadian heritage with the rest of the biological world. Like all living organisms, we have an internal clock that sets the timing of our daily behavior and physiology. Also shared is our sensitivity to the environmental fluctuations of light and darkness.

Chronobiology seeks to harness this new knowledge to understand the "when" and "why" of human behavior. Why does someone become depressed at this particular time of year? Why does one's attention flag every afternoon at the same time? Why can't our children go to sleep until well after midnight? Chronotherapy picks up the baton, seeking to provide rhythm-changing interventions that address these problems. And there is no one who has led the development of this field more than psychologist Michael Terman of Columbia University.

Beginning with his early research on the role and properties of the eye in the circadian system, through his elegant demonstration of the antidepressant effect of bright light therapy for seasonal affective disorder, and on to his more recent studies of low intensity, progressive lighting administered at the end of the sleep cycle—Dr. Terman's group has been at the forefront of this creative new discipline. Like any true pioneer, he has insisted on exploring the possible range of light-based treatments, examining its potential utility in nonseasonal depressions and bipolar disorder, childhood depression, attention deficit hyperactivity disorder, Parkinson's disease, and dementias. As an innovative chronotherapist, Dr. Terman has also championed the use of sleep manipulations and carefully timed melatonin usage to shift rhythms and enhance daytime alertness and behavior.

In this book, his first for a general audience, Dr. Terman presents his hard-won discoveries in clear, easy-to-understand language. This translation from the technical, scientific literature is expertly achieved in collaboration with Dr. Ian McMahan, who is well known for his work and writings in developmental psychology. Using real-world examples, they introduce us to people with depression, insomnia, seasonal affective disorder, and jet lag and shift work disturbance. We learn of the challenges and difficulties that make up each of these conditions. We are then helped to understand these problems from a chronobiologist's perspective: What is taking place with the nature and timing of circadian rhythms in each of these problems? Is internal timing falling out of synchrony with the light/darkness cycle? Are certain internal rhythms becoming misaligned with each other? Finally, we are taught about the variety of chronotherapeutic approaches that can be used to correct these timing disturbances. From the now-mainstream use of bright light therapy, through explanations of melatonin usage, sleep timing modifications, and dawn simulation, the full range of rhythm-shifting strategies is laid out in basic and elegant fashion.

In focusing on rhythms and timing, chronobiology offers a new way to understand and approach the wide range of behavioral problems that afflict so many of us. This new discipline moves us beyond an exclusive focus on neurotransmitter dysfunction and the corresponding reliance on psychotropic medications. We all benefit from this expansion. This book points us toward a new and enriched age of mental health care.

Preface

We are all animals! No put-down meant there. On the contrary, our shared biology with the animal kingdom offers powerful insights into not only obvious disease processes, but also our own inner world. We can never truly know what an animal "feels," although we can often guess from things like purring, tail wagging, and crying. Despite the limitations, we have learned much from animals about essential psychological mechanisms—our well-being, or lack of it.

Years ago, at Brown University in Providence, Rhode Island, Michael and his research group were investigating how brain circuits affected what looked like "pleasure" in lab rats. When stimulated directly with tiny electrical pulses deep inside the brain, the critters would incessantly "ask for more" by returning to the switch in their testing chamber that activated the pulses. When we allowed them to live twenty-four hours a day in the chamber, with plenty of food and water, they self-stimulated to the point where they ignored eating and sleeping.

It was like an unceasing addiction to pleasure. Except for one thing: After several days of nonstop indulgence, the self-stimulation started gradually to rise and fall each day even though there were absolutely no day/night cues in the soundproof, darkened chamber. At the peak of the cycle, an animal might compulsively click the switch ten thousand times an hour, taking no time to rest or eat, but occasionally running to the water bottle for a few sips. At the midpoint of the cycle about six hours later, the animal might click in bursts up to five thousand times an hour, interspersed with eating, drinking, grooming, and brief pauses. Even at the nadir of the cycle, spurts of self-stimulation persisted, but only up to about one thousand clicks an hour, interspersed with ten- to twenty-minute bouts of sleep, as is typical for lab rats. This oscillation of behavior—from peak to midpoint to nadir to midpoint—persisted over many weeks, and the animals remained healthy, and apparently quite happy.

But with such a long stream of data, we noticed that the ups and downs in self-stimulation were coming about thirty minutes later with each passing day. If the cycles represented a day length for these animals, it was not the twenty-four-hour solar day, but something longer—like what one might see on another planet with a slower rotation speed. So we imposed a solar day by turning the lights on and off every twelve hours, and within a few days the pleasure cycle exactly matched the day/night cycle, with the peak at night and the nadir during the day. (Rats, you'll remember, are nocturnal animals.)

How was this working? Was the light merely defining the solar day, and the animals reacting accordingly? No. When we left the lights on continuously—again removing the solar cycle—the oscillation got even longer than it was under constant darkness, with ups and downs about every twenty-five hours. The mystery was solved when we presented a single fifteen-minute pulse of light in the otherwise darkened chamber.

When we gave the pulse while self-stimulation was on the rise, the next day's frenetic activity shifted markedly later. When we did this while self-stimulation was on the decline toward the animal's quiet period, the next day's activity began earlier than expected.

The light, we concluded, must be signaling an internal clock that operates on a time base different from twenty-four hours, causing it to reset earlier or later. Take an animal whose self-stimulation in darkness cycles at twenty-five hours. With only a brief light pulse every morning, the oscillation could shift an hour earlier—and the result would be an exact twenty-four-hour rhythm! Our results fit with reports by other investigators who had observed non-twenty-four-hour animal activity cycles, and adjustment to day/night cycles, twenty years earlier. While they had analyzed running wheel activity, feeding and drinking, we had extended it to the pleasure principle (at least in terms of the rat's self-stimulation). The timing of a light pulse resets the clock, but the clockwork is internalized.

Around the same time—which is quite a while ago, in the early 1980s—scientists at the National Institute of Mental Health discovered that bright, indoor light exposure could make clinically depressed, oversleeping patients better within a week's time. Might light therapy be signaling the human's internal clock in a similar way as it seemed to be doing with lab rats? Michael, the basic researcher, got so excited about this possibility that he switched fields—from animal physiology and behavior into psychiatry. Along with many close colleagues around the world, he and his research group spent the next two decades in clinical trials of light therapy for depression. They were able to verify the suspicion that dysregulation of internal clockwork is close to the heart of depressive and sleep disorders, and that an effective way to treat these problems is to restructure how we expose ourselves to light. With this book, we hope that patients, family members, and doctors will taste the depth and breadth of this rather straightforward discovery.

We are grateful to many colleagues who collaborated in research and offered their insights for our book: David Avery, M.D., Francesco Benedetti, M.D., Randall Flory, Ph.D., John Gottlieb, M.D., Markus Haberstroh, Dipl. Arch. FH, Charlotte Remé, M.D., Paul Riccobono, president, Complete Control Solutions, Inc., Norman Rosenthal, M.D., Elizabeth Saenger, Ph.D., Dorothy Sit, M.D., Oliver Stefani, Dipl. Ing., Jonathan Slater, M.D., Harm-Pieter Spaans, M.D., Robert Spitzer, M.D., Jonathan Stewart, M.D., Gregory Sullivan, M.D., Thomas White, M.D., Janet Williams, D.S.W., Anna Wirz-Justice, Ph.D., and Katherine Wisner, M.D. And Michael expresses deepest appreciation to his lab partner since 1964, Jiuan Su Terman, Ph.D., and their professor at Brown University, Julius Kling, Ph.D.

All thanks to the anonymous chronotherapy patients (plus one mom), shift workers, and jet lag and time zone sufferers who lent their personal stories for our edification and our readers' benefit. We faced an ethics issue reporting on the experience of patients seen at our Columbia clinic. Their privacy could be assured by changing names, locations, and other potential personal identifiers. They offered their consent and were not paid for participation. But Michael did not want to bias their stories by interviewing them himself. So instead, we enlisted the collaboration of clinical psychologist Margaret Mandel, Ed.D., whose expert, sensitive interviewing produced a set of truly balanced case reports that address the successes as well as the difficulties in using chronotherapy.

We wish to acknowledge our associates who helped so much to bring this work to publication: our agent, Bob Mecoy of Creative Book Services; Lisa Johnson, Casey Maloney, Megan Newman, Miriam Rich, and our perspicacious editor, Marisa Vigilante of the U.S. Penguin Group; and Nikki Hafezi of the Center for Environmental Therapeutics. For their critical reading and smart advice, our thanks go to Emily DeCola, Kate Lardner, and Jane and Selena McMahan. Stephen Scoble was our generous, unflappable portrait photographer.

Michael's clinical research has been supported by the Center for Environmental Therapeutics, the Irving Institute for Clinical and Translational Research at Columbia University Medical Center, the National Institute of Mental Health, the New York State Office of Mental Health (New York State Psychiatric Institute Division), and the New York State Science and Technology Foundation. Neither the publisher nor the authors have financial interest in the products described herein.

MICHAEL TERMAN

IAN MCMAHAN

New York City

Introduction

In a New Light

Our lives are shaped by the spaces we live in and move through—the rooms and streets, the hills and valleys, the cities and deserts. They are shaped just as powerfully by our engagement with time. Yesterdays and tomorrows may seem to stretch in an unending line, but within each day is a cycle that is echoed by the ebb and flow of our inner life—our attention, alertness, energy, and mood. This internal rhythm is produced and regulated by a neurological mechanism called the *circadian clock*, and we are all born with it. The crucial ways it affects our adjustment to internal and external reality is the focus of the new scientific field called *chronobiology.*

The circadian clock, which has been traced to a tiny section of the brain, is intimately connected to both our internal body systems and our external environment. These connections work in both directions. The clock in the brain, guided by our genetic heritage, makes us wakeful, sleepy, mentally alert, physically vigorous, happy or sad, at certain times of day. These rhythms guide the way we interact with the external world—when we get up, go outside, and go to bed. In turn, the daily pattern of light and darkness we encounter affects the settings of the internal clock.

When this feedback loop runs smoothly, life is good. But sometimes these adjustments can go awry. Illness, genetic bad luck, the circle of the seasons, man-made changes such as dark apartments and odd work hours can all potentially disrupt the close link between inner clock and outer world. When this happens, problems develop that affect aspects of our lives ranging from attention and concentration to emotions and energy.

Sleeping and Waking

The inner clock is vitally involved in guiding our patterns of sleeping and waking. When it gets out of sync with the external day/night cycle, we can encounter a variety of difficulties:

- We may find it impossible to fall asleep when we think we should, or blame ourselves when we can't get up on time.
- We may fall asleep well before a reasonable bedtime, then find ourselves up and about in the middle of the night, praying for the day to start.
- We may live under the illusion that "If I could just get more sleep, I'd have enough energy to get through the day," then find that we are wiped out after sleeping in.
- We may try, but fail, to get to sleep with drugs while we are still wide-awake. When we plead for help, our doctor keeps increasing the dose or juggling medications without results.

Worse still, many people with sleep problems come to feel they must have a mental or moral failing. They think, "If only I would buckle down and exert my willpower!" The feeling of isolation deepens as family members chide them for agonizing over something as simple and everyday as getting to sleep.

Millions of people have a circadian cycle that makes it hard to fall asleep before midnight or later. Sleeping unusual hours does not necessarily spell trouble, though. This is not unnatural or abnormal—it's simply different from the majority. It only becomes a problem if a late sleeper is forced to conform to the majority's nine-to-five workday schedule. Many of these night owls come to terms with their personal reality and even structure their lives around it. For example, you are unlikely to have much success as a stage actor or jazz musician if you always fall asleep by 11 PM.

Although circadian timing is at the heart of common sleep problems, it is one factor among many. The sleep centers of the brain are sensitive to influences that range from psychological and biological to environmental and pharmaceutical. Dealing with a sleep problem calls for close attention to the person's individual mix of causes. But because the circadian clock is so important in determining the sleep/wake cycle, we believe that the first step should be to carefully examine a person's rhythm, and, if indicated, zero in on rhythm adjustment before appealing to other therapies.

Mood and the Inner Clock

Tens of millions of Americans struggle with mood problems that include prolonged sadness and depression. Among them, a great many grapple with both mood *and* sleep problems, each feeding into and aggravating the other. In fact, changes in sleep patterns are among the major symptoms doctors use to diagnose clinical depression. Often, when the sleep problem is treated directly, the mood problem resolves itself. This is one indication that the inner clock is deeply involved in determining our daily moods. It's a two-way street: Depression can trigger sleep problems, whether insomnia or oversleeping. And sleep problems can trigger depression, and sometimes even *predict* the onset of a depressive episode.

As many as one out of three Americans experience winter slumps, periods of gloom and despair that come on during the fall or winter and typically last until May. In its more severe form, this is seasonal affective disorder, or *SAD*. Research in our clinic and many others has shown that SAD is directly linked to the circadian clock and the way it is affected by the daily light/dark cycle. Understandably, people who suffer from SAD develop strong intuitions about the cause of their recurring cycles: an intensified winter work schedule, the pressure of the school year, past life events that happened to occur in winter, such as deaths in the family or family stresses at required holiday reunions. What these explanations tend to overlook is the impact those long winter nights and all-too-brief winter days have on the circadian clock.

If light and darkness are such strong mood regulators, should we not be zeroing in on environmental causes of depression before appealing to other therapies?

Light and Dark

The changing seasons are not the only reason the inner clock may start to present problems. We may not get enough light even when it *is* available outdoors. We may get light, but at the wrong time of day (or at night)—especially artificial indoor light. One of the main functions of the inner clock is to signal the pineal gland deep in the brain to produce the night-time hormone, melatonin, which facilitates sleep. But with the wrong lighting pattern, the pineal gland may turn off or begin producing melatonin at times we want to be awake. Or we may have circadian clocks that put our physiology at odds with external time. Any of these factors can have major consequences, including gloom and despair, sluggishness, difficulty working, and a sense of pervasive anxiety.

In contemporary life, the inner clock faces challenges beyond the predictable cycling of day and night. Consider the strange experience of quickly crossing multiple time zones and having to switch to a new day/night schedule when the body still thinks it is at home. This is the dreaded jet lag, with its disorientation, fatigue, and waking up in the middle of the night. And closer to home, the pressure by industry to produce nonstop forces millions of shift workers to stay awake and alert while out of sync with their circadian clock. This arbitrary disruption of their lives can have terrible consequences both for them and for their families.

How We Live Today

Most of us live a twilight existence. We spend most of the day indoors, with light levels that are a mere fraction of daylight intensities. The amount of time we are likely to spend outdoors with sufficient exposure to light is trivial. Come night, the situation is reversed. We are flooded with light pollution from televisions and computer screens. Even after we turn off the electronics and go to bed, there are streetlights, neighbors' outdoor flood lamps, the headlights of passing cars, and even self-imposed night-lights. These directly disturb the quality of our sleep and provoke mini–jet lags by affecting the circadian clock during its most sensitive interval.

Homes have evolved over the millennia to protect us from the elements and from extremes of heat and cold. The amount and quality of light we receive indoors has been given much lower priority than protection from the weather. Neither architects nor the public have yet discovered the pervasive biological role of light in optimizing sleep and daytime well-being. The misguided indoor lighting that results contributes greatly to insomnia, depression, poor concentration, and fatigue.

From Before Birth Onward

Circadian rhythms are a basic fact of life, throughout life. However, at different points in the life span, the particularities of their influence change, sometimes dramatically.

Pregnancy is a time when many women feel their sleep and mood go awry. However, they may shy away from treating these problems with medications that can pass to the fetus, with consequences that are still incompletely understood. There is also growing evidence that the child's circadian system begins to be shaped by environmental influences even before birth. Then comes the vulnerable postnatal period when all new parents are worried about lack of sleep. The baby's eyes start to send messages about light and dark to the inner clock. Feeding and sleeping patterns go from scattered to consolidated as the baby adjusts to the day/night cycle in its new environment outside the dark womb. During childhood, the need for lots of sleep can come into conflict with bright, noisy home environments and other interferences, leading to long-term emotional and sleep problems.

Anywhere from one-third to three-quarters of teens suffer from some kind of sleep disturbance, and about 20 percent of teens will go through a bout of depression before they reach adulthood. A major reason for these problems is a new conflict between the teen's inner and outer schedules. At puberty the inner clock changes its settings. Before, it may have signaled that it is time to sleep at 9 PM and time to wake up at 7 AM. Now it shifts in the direction of signaling sleep at midnight or 1 AM and waking up at 9 AM or 10 AM. However, the teen's social environment—family, work, and especially school—does not accommodate this shift.

Most teens move toward a more adult sleep pattern during their twenties, but for some, the biological urge to stay up late and sleep late continues into their forties. This is a time when learning how to use the light/dark cycle can bring powerful benefits.

As we age, the character of our sleep problems tends to change. Many elderly people start to become deeply drowsy during the afternoon. By 8 PM, after dinner, they are likely to fall sleep (often in front of the TV), only to find themselves awake and alert at 3 AM, the "hour of the wolf." This bizarre and disturbing pattern is linked both to biological changes with age and to behavioral responses to these changes. With age, the functioning of the circadian clock in many people begins to weaken and the pineal gland produces less melatonin. This reduces the physiological signal of darkness timed by the inner clock and makes it harder for the body to switch to night-appropriate functioning.

Resetting Your Inner Clock

Understanding the reasons why we may have sleep or mood problems is an excellent thing. Better still is also knowing how to deal with the problem or even to get rid of it. The point we most want to emphasize in this book is that these widespread problems of sleep and mood—across occupations, lifestyles, and life span—are in major part caused by environmental factors *that we can control*. This insight is the fruit of laboratory and clinical research that began in the 1980s and continues at an accelerating pace. As a result, chronobiologists have learned how to *reset* the inner clock when it gets out of sync with our external environment. With this understanding, we have new, practical ways to improve our sleep, mood, energy, and overall quality of life. And often we can do it naturally, without the need for medications whose mechanisms of action and long-term effects on the body and nervous system are still poorly understood.

Our evolution and the success of our species have taken place in the overarching context of our natural environment. But in the last couple of hundred years, the industrial and post-industrial eras have confounded the order of things. The result is an unnecessary burden of medical and psychological illness. Of course modern technology and modern medicine have made our lives incomparably better in so many ways. At the same time, however, they have cut us off from many aspects of our physical world that our bodies and brains evolved to rely on. To correct this, we have to identify and revive the positive elements of the pre-industrial environment of our forebears—starting with light and air.

Science is finding out a tremendous amount about the neurochemical processes in the brain that regulate sleep and mood. One predictable consequence of this progress is a clinical focus on developing and using drugs, such as sleeping pills and antidepressants, which intervene in these chemical processes. But when we use modern methods of imaging to examine brain processing, we find that the impact of these drugs on target regions looks remarkably like the action of the environment!

We are near the point of making a bold assertion: These drugs work because they happen to *mimic* environmental influences that achieve the same effects. But there is one very significant difference between drugs and environmental influences. The drugs bathe the brain and body with chemicals that have multiple, often unknown effects and possibly dangerous side effects. By contrast, the parallel environmental factors—especially light, darkness, and timed wakefulness—can act *specifically* to achieve the desired therapeutic responses.

The two environmental factors that we focus on are both vitally important and incredibly simple:

- *When does the sun come up?* We mean this both literally (Where do you live? Is it winter or summer?) and figuratively (When do you get outdoors? For how long? How bright is the lighting in your home and workplace? When do you turn the lights up and down?).

- *How is your air quality?* The air we breathe indoors and on city streets does not match the air at the seashore or in the rain forest. We sleep and feel better when we breathe fresher air—and we don't mean air-conditioned air!

One answer to the predicaments of modern urban life would be to move to a sunny tropical clime, preferably next to a beach or waterfall. Some have the liberty and means to do just that. For the rest of us, we are lucky to have access to effective new technologies that offer at least some of the same advantages as that beachfront cottage, at a far more affordable cost in time, money, and convenience. And best of all, they are scientifically based, clinically proven, and environmentally sound.

Getting Off to a Faster Start

Naturally, we hope you will read, enjoy, and benefit from all of the information in this book. The best way to understand chronobiology and chronotherapy is to read the entire book, but we also realize that you or someone you care about may have a particular concern that you are looking for help in dealing with. In that case, you may want to go straight to the chapter that most directly addresses your concern. To help you do that, here is a series of questions you can ask yourself:

1. You really try to keep to a normal workday schedule, but . . .

☐ You can't fall asleep on time.

☐ You have tried sleeping pills or sleep aids, but you are not happy with the results.

- ☐ You use the snooze button to avoid getting up.
- ☐ You oversleep—either every day or on weekends and holidays.
- ☐ You feel most alert after midnight.

IF YOU answered *yes* to some of these questions, it sounds as if you may be experiencing problems with *delayed sleep phase disorder*. Chapter 3 goes into details.

2. For days or weeks at a time . . .

- ☐ You get so involved in activities that you forget about meals—and you don't miss having them.
- ☐ You start interrupting people in conversation, or telling bad jokes, or arguing too much.
- ☐ You feel like you're "flying," both mentally and emotionally.
- ☐ Sleep becomes secondary—most of the time you're wide-awake.
- ☐ You're at risk for getting into real trouble because of your behavior related to speeding, spending, fighting, drug use, or sex.

IF YOU answered *yes* to some of these questions, it sounds as if you may have experienced problems with *bipolar disorder*, which may need the attention of a specialist. Chapter 5 goes into details.

3. About depression and light availability . . .

- ☐ You take one or more antidepressants prescribed by your primary care doctor or psychiatrist.
- ☐ You live in a dark house or apartment or your bedroom is pretty dark when you wake up—or both.
- ☐ You work in a windowless space, or your desk/workbench is far from any windows.

☐ On average, you are not outside during the day for more than fifteen to twenty minutes.

IF YOU answered *yes* to some of these questions, it sounds like you may be suffering from depressed mood as the inadvertent result of insufficient light exposure. Chapter 5 goes into details.

4. In the winter . . .

☐ You develop unhealthy food cravings and start to gain weight.
☐ You start sleeping much longer than in summer.
☐ You feel fatigued during the day, while in summer you're vigorous.
☐ You start cutting yourself off from friends, family, and coworkers.
☐ You start going for sweets, especially in the second half of the day.

IF YOU answered *yes* to some of these questions, it sounds as if you may have experienced problems with *seasonal affective disorder*. Chapter 5 goes into details.

5. For weeks or months at a time . . .

☐ Your appetite increases or decreases noticeably.
☐ You feel more tired.
☐ Things that you normally find involving, including work, lose their interest.
☐ You look and feel like you're "down in the dumps."
☐ Your sleep takes a turn for the worse.

IF YOU answered *yes* to some of these questions, it sounds as if you may have experienced problems with *clinical depression*. Chapter 9 goes into details.

6. About air . . .

- ☐ You feel better as soon as you get to the country.
- ☐ You look forward to taking a walk as soon as a spring thunderstorm has passed.
- ☐ The air circulation at home feels stuffy or stale, even if the air conditioner is on.
- ☐ Allergies keep you awake at night, so you're tired during the day. Allergy drugs make you drowsy or "out of it."
- ☐ You haven't responded well to antidepressant drugs, and want to try a recently discovered natural alternative.

IF YOU answered *yes* to some of these questions, it sounds like you may want to try improving your indoor air quality with high-density negative air ionization. Chapter 10 goes into details.

7. You've given birth within the past year.

- ☐ You find yourself napping during the day.
- ☐ You are getting up to care for the baby in the middle of the night.
- ☐ You're feeling irritable with family members, including the baby.
- ☐ Despite fatigue, it's hard to fall asleep at bedtime.
- ☐ You wish your partner would or could help more with your childcare responsibilities.

IF YOU answered *yes* to some of these questions, you may be grappling with postpartum depression, which is experienced by as many as 40 percent of new mothers. Chapter 11 offers ways to relieve this problem.

8. You are a teen or young adult.

- ☐ You can't concentrate in class in the morning (or, if you're in college, you have arranged for late classes).
- ☐ You don't have time for breakfast, or you wake up without any appetite.
- ☐ You sleep very late on weekends.
- ☐ You text, tweet, or go on Facebook late at night.
- ☐ You're glued to your PC or tablet screen doing homework until right before bed.

IF YOU answered *yes* to some of these questions, and have friends who might answer the same way, take a look at Chapter 13 for our suggestions to improve the picture and get out of the trap.

9. You have to go out to work at night and . . .

- ☐ Around 4 AM you have to force yourself to keep your eyes open.
- ☐ You are tanking up on coffee or sweet carbs (like doughnuts).
- ☐ You dread the sensation of going home while the sun is rising or already high in the sky.
- ☐ Even though you're sleep deprived, you can't get satisfactory sleep during the day.
- ☐ You're having family stresses because you're not around enough.

IF YOU answered *yes* to some of these questions, it sounds like you may have experienced problems with shift work disorder. Chapter 15 goes into details.

10. After flying across time zones for work or vacation . . .

- ☐ You feel physically clumsy.
- ☐ You get light-headed or dizzy.
- ☐ You suddenly get sleepy in the middle of the day.
- ☐ You can't think clearly or keep concentration.
- ☐ You suddenly wake up in the middle of the night.

IF YOU answered *yes* to some of these questions, it sounds like you may have experienced problems with *jet lag*. Chapter 16 goes into details.

11. Whatever specific concerns you have . . .

- ☐ You like to keep up with the latest scientific developments.
- ☐ You want to know *why* you're having sleep or mood problems, not just how to treat them.
- ☐ Your doctor hasn't heard about chronotherapy, and you want to help him or her learn (and pay attention).
- ☐ You're surprised to find that timed daily light exposure can act like a drug—sometimes better!
- ☐ You want to convince someone that there's real substance to chronotherapy.

IF YOU answered *yes* to some of these questions, read the book from the beginning!

PART 1

Time, Sleep, and Rhythms

1.
External Time vs. Internal Time

A plant called *Mimosa pudica*, or "touch-me-not," has a very odd habit. Toward the end of each day, as night approaches, the leaves fold inward and close up. And every morning near dawn, the leaves fold out again. In the early eighteenth century, a French scientist with the resplendent name of Jean-Jacques d'Ortous de Mairan noticed this and found it deeply intriguing. His first thought was that the leaves were somehow sensitive to light and darkness. As the level of light changed at dawn and again at dusk, the leaves must be reacting. This explanation no doubt seemed obvious, but de Mairan wondered if it was scientifically correct.

To find out, he performed a simple experiment. He put mimosa plants inside a light-tight box and watched to see what would happen. To his surprise, the leaves went on opening near dawn and closing near sunset even though the plants were in continuous darkness. Did each plant contain its own internal clock that kept working whether or not it received sunlight? De Mairan found this idea so disturbing that he decided not to publish the results of his experiment. A friend had to do it for him. But in the years that followed, one scientist after another repeated de Mairan's experiment and got the same result. The mimosa had its own clock. And if one kind of plant had an inner clock, why not others? Why not animals? Or even humans?

In one way, it is easy to see why people a couple of centuries ago, or even today, might find this idea upsetting. An internal clock, constantly ticking away and governing what we do? It calls forth the image of an *automaton*, one of those spooky clockwork chess-playing machines that eighteenth-century Europe found so fascinating but ominous. Or in today's terms, a robot—a gadget that threatens to blur the crucial boundary between living and non-living. Would that leave any room for free will and self-determination?

On the other hand, if there are internal clocks, it really should not come as such a big surprise. One of the most obvious ways we understand and experience time is based on the earth's twenty-four-hour cycle of rotation, or *solar time*. Depending on the time of year and the distance we are from the equator, the relative length of day and night vary, but together day and night always add up to twenty-four hours, plus or minus a few seconds. Over the aeons, every living thing on Earth, humans included, has adapted to and internalized that basic fact of life.

Internalized, yes—but not in a form that exactly matched the solar cycle. A hundred years after de Mairan's discovery, a Swiss botanist named Augustin de Candolle tried keeping mimosas in darkness over a number of days. Even without the time cues of dawn and dusk, the plants continued to show their daily pattern of opening and closing, just as they had for de Mairan. But they also did something even odder. Each day their leaves opened and closed an hour or more *earlier.* Gradually, their timing drifted more and more out of agreement with solar time. Deprived of light cues, the plants were following a daily cycle of twenty-two to twenty-three hours instead of the familiar twenty-four-hour solar cycle. So a plant's internal clock was not only independent of light levels, it could also be at odds with the familiar, fundamental twenty-four-hour day. To use a more recently adopted term, the darkness allowed the inner clock to *free run*; in other words, it followed its own internal rhythm rather than staying tied to the solar day.

Do animals, like plants, have a built-in way of telling time? Certainly, it is easy enough to show that many animals follow daily rhythms. In our Preface, we described how we discovered such rhythms in lab rats by letting them live in a chamber where they could self-stimulate the pleasure centers of their brain. The pattern gradually cycled up and down each day.

But you do not need a laboratory to find this out. Anyone who has kept a pet hamster knows that it loves to run in its exercise wheel at night, usually just when you would like to get to sleep. And if you put the hamster in continuous darkness for a few days, it will still go on running during the nighttime hours. Why? Clearly it is possible that the hamster has an intrinsic inner clock. But there are other possibilities, too. Its daily rhythms might somehow be following external changes *other than* light, such as air temperature or noise levels. Maybe the lab inadvertently set the rhythm through the timing of meals. Or maybe something still subtler is involved, such as tiny alterations in the earth's magnetic field as it rotates on its axis. These were some of the questions twentieth-century scientists ran into as they began to explore the inner clocks of animals.

Many of the earliest findings were about insects. A Swiss psychiatrist whose hobby was entomology was on summer vacation in the Alps. He noticed that bees showed up on his terrace each morning during breakfast. They were such a nuisance that the family moved breakfast indoors. But the bees continued to arrive on the terrace at the same time, even though there was no more marmalade there to attract them. A later researcher deliberately gave bees sugar water at the same time for several days, then eliminated the sugar water and watched to see what would happen. The bees arrived at the usual time. Even when he tried this experiment again deep in a salt mine, far from light or temperature cues, the bees somehow knew when it was their habitual feeding time. Most striking, perhaps, a French researcher trained bees to find sugar water during a specific two-hour period each evening. He then took them on an overnight flight to New York, where a similar lab was waiting. Once again the bees came looking for food during the usual two hours—*French* time!

During the 1950s, German physiologist Jürgen Aschoff began recording the daily rhythms of birds and mice when they were kept in constant conditions. Like others doing similar research, he came to the conclusion that biological processes such as sleep, activity, and body temperature are controlled by an innate inner clock that functions even when it is deprived of external cues. He extended this work to humans, building an underground bunker to isolate his volunteers from external light and temperature changes for several days at a time. Important as the research was, this was not what captured the imagination of the public and made them aware of daily rhythms and inner clocks.

In the summer of 1962, Michel Siffre, a twenty-three-year-old French geologist and passionate spelunker, set up camp in a glacial cave 250 feet below the rugged mountainous terrain near the French-Italian border. His equipment included a small red tent, a down sleeping bag, some books, and a weak flashlight. Most important, he had a field telephone linked to a team of record keepers on the surface. What he did *not* have was a watch, clock, or any other way to keep accurate track of the passing hours and days. His goal was to spend two months "beyond time," isolated from any information about the outside world. This was an era when space explorers were first orbiting the earth, when sailors in nuclear submarines were starting to cruise submerged for weeks at a time, and when ordinary people, frightened by Cold War tensions, were wondering if they would have to finish their lives in an underground shelter. Siffre hoped that whatever he learned in his months in the cave would be useful in all of these situations.

Each time Siffre woke up, ate a meal, or got ready to sleep, he sent a signal to the observers up above. In between, he explored the farther reaches of the cave, took geological samples, and practiced climbing almost sheer ice sheets. Conditions in the cave were harsh, with 98 percent humidity and temperatures constantly below freezing. His tent, sleeping bag, and down-filled booties quickly became sodden. He had a small stove, but he didn't dare leave it going while he slept, for fear of asphyxiating himself. At one point his body temperature went down to 34°C (93°F). But none of that was as bad as the sense of floating in an unknown space. "Time passed without my being aware of it in the darkness and silence," he recalled. "I felt I was on another planet. For the most part I dwelt neither in the past nor in the future, but in the hostile present. In that environment everything was against me: the rocks that crashed down from time to time, the damp, chill atmosphere, the darkness. I waged a ceaseless combat with myself to surmount these terrors by sheer willpower."

Waking up in total darkness was his worst moment. How long had he slept? If he felt rested, he guessed ten hours; if he was tired, he guessed two hours. But whichever he guessed, it never seemed to fit well with what he thought the day and time might be.

On September 14, as prearranged, Siffre's friends called him to get ready to leave the cave. He was sure they were trying to fool him into quitting early. According to the journal he had been keeping, it was only August 20! Later, after he recovered from his experience, he compared his journal entries with the detailed records his friends had kept. One time, when he had thought he had spent a few boring hours doing chores and working on his journal, he was actually awake for eighteen hours straight. Another time, what he had thought was a two-hour nap was in fact a full ten hours of sleep. But like the plants, insects, birds, and mice studied by earlier researchers, over the weeks Siffre showed a very regular pattern of waking and sleeping times. Together, they consistently added up to twenty-four hours and thirty minutes. Like de Candolle's mimosas, Siffre had an inner clock with a daily cycle that differed from the twenty-four-hour solar cycle. Deprived of external time cues, his inner clock had started to free run. It kept accurate time for his own private cycle but each day cast him more and more adrift from the world outside.

This wasn't Siffre's only long-term cave adventure. In 1972 he spent six months in Midnight Cave, near Del Rio, Texas. This time he was warm, well fed on astronaut rations, and aware that his perception of passing time was almost certainly inaccurate. Even so, his inner clock played some odd tricks on him. Twice, his sleep-wake cycle even doubled in length, to almost fifty hours, leading him to stay awake and active for up to thirty-six hours, followed by twelve to fourteen hours of sleep. He was aware of this only because he got so much more work done during those "days." When he returned to the surface, his estimate of the date was two months off. And in late 1999, Siffre, now almost sixty-one, stayed for two months in the Clamouse Cave in southern France. He planned to welcome the new millennium on New Year's Day with champagne and foie gras, three thousand feet underground. Once again his inner clock fooled him. He was three and a half days late.

Siffre's exploit in the early 1960s helped draw more scientific attention to what were now beginning to be called *circadian rhythms*. This term was based on the Latin words *circa* (around or about) and *dies* (day) and can be understood as meaning both "around the twenty-four-hour day" and "about twenty-four hours." But instead of hunting for suitable caves, researchers built isolation rooms, whether underground or not. These were designed to eliminate any changes in light, sound, and temperature that might serve as a clue to the passing of time in the world outside.

Today, a temporal isolation lab looks very much like a comfortable motel room, but with some noticeable differences. No windows, no clocks, no real-time TV or Internet access. In one corner, a kitchenette, in another, exercise equipment. A desk, table, and bed complete the furnishings. There, volunteers can spend several weeks cut off from outside time cues. They go to sleep when they feel sleepy, wake up whenever they feel like it, fix a meal or snack when they're hungry—and, one hopes, get in some exercise now and then. Meanwhile, sophisticated instruments track their sleeping and waking, changes in body temperature, activity level, blood factors, and more.

Who signs up to take part in such an experiment? It might be a graduate student writing a dissertation, a cellist who wants to avoid all distractions and concentrate on practicing intensively for an upcoming performance, or in fact anyone who is healthy and curious, has no other time commitments, and can use the cash reward for participation. And what is it like for them? On Night One, our graduate student, call him Bob, goes to sleep at 11 PM, the way he usually does, and wakes up at seven the next morning. And as far as he knows, each of the subsequent days follows the same pattern. He is confident that he is living a twenty-four-hour day and has no reason to think otherwise.

The instruments in the other room tell a different tale. On Night Two, Bob in fact goes to sleep at 11:30 PM and wakes at 7:30 AM. Night Three sees his bedtime pushed to midnight, Night Four to 12:30 AM, and so on. His circadian cycle, as dictated by his internal clock, turns out to be twenty-four and a half hours long, though in different individuals it might have been anywhere between about 23.9 hours and 26 hours long. After ten days of this free running (free of external time cues), he's going to sleep at 4 AM. In the outside world dawn is about to break, but this has now become the start of his internal *night*, which will last right through the morning hours. When his stint in the isolation room ends, he sees a clock for the first time in three weeks. He cannot believe it: His subjective time is a full half-day off.

During the next three weeks, another volunteer, Peggy, takes Bob's place in the room. She is more of an "owl" than Bob. She stays up until 1 AM on Night One and sleeps until 10 AM. Like Bob, she is convinced that she is simply following her normal schedule. But this time the gap between subjective and objective time widens faster. On Night Two, her actual bedtime is 2 AM, and on Night Three, 3 AM. Without knowing it, Peggy is losing an entire hour a day. Over the course of three weeks, her sleep-wake pattern marches all the way around the clock, almost to the point where it started. Her internal clock tells her that it is Thursday lunchtime, but in fact it is Friday morning.

Research of this sort has uncovered a great deal about the internal sleep/wake cycle. First, its length varies tremendously from one person to the next. For the largest number of adults, the cycle is roughly twenty-four hours and twenty minutes, but some have cycles very close to twenty-four hours (though almost always at least a few minutes longer). Others have much longer cycles, up to twenty-five and a half hours or even longer. There are even a few rare individuals whose clocks cycle at *less* than twenty-four hours.

There are still many mysteries about the circadian cycle. For example, what was it that caused Michel Siffre and other temporal isolation participants to sometimes switch from a twenty-four-hour cycle to a forty-eight-hour cycle? One thing that is becoming clearer, though, is that the individual cycle length is largely set by genetics. Researchers are even closing in on particular genes that help determine the circadian rhythm in different species. The first of these to be identified in mammals was the CKI epsilon gene in golden hamsters. This gene occurs both in its normal form and in what is called the tau mutation. Most hamsters—those with two of the normal gene—have a circadian rhythm with a period of about twenty-four hours. In those that carry one normal and one mutant tau gene, the pace speeds up to a period of twenty-two hours. As for those with two tau genes, they show an amazingly fast circadian period of twenty hours. In isolation, one of these mutant hamsters would complete a week of daily cycles in less than six calendar days!

Specific genes also affect the circadian pattern in humans. For example, researchers have studied families in which many members have an unusually early sleep cycle. Those with this condition, called Familial Advanced Sleep Phase Syndrome (FASPS), have sleep-wake and body temperature rhythms that in isolation cycle in as little as twenty-three and a third hours. Genetic mapping has revealed that most people with FASPS carry a particular form of a gene called hPER2. This is an instance of nature's stinginess, because hPER2 turns out to be a human version of the same gene that governs the circadian clock in fruit flies.

We need to remind ourselves at this point (and even more frequently) that genetics is *not* destiny. The way someone's genes get expressed in real life, the person's *phenotype*, is affected by a host of factors, both singly and in combination, from the moment of conception or even earlier. A child born in a famine-stricken region may have the genetic makeup of a future basketball star, but even if he survives to adulthood, he is not likely to realize his genetic potential. In the case of FASPS, it turns out that in one affected family, a few of the members have just as much of a sleep problem as their relatives, but they do not carry the mutated version of hPER2. Their phase-advanced sleep apparently has no genetic basis at all—it's a matter of being in close contact with those who *do* have the mutation, and the likelihood that the whole family has been living on a light/dark schedule far earlier than local time.

Whatever our genetic makeup, our individual circadian cycles also change as we move from birth to old age. In Chapter 11, we look at recent evidence that the circadian timing system starts to develop even before birth. However, babies generally do not show strong rhythms in sleeping and waking or in hormone secretions until about two months. Starting at puberty, the internal clock tends to slow down. It levels off in the early twenties and does not get back to a "normal" speed until early middle age. Then as we continue to age, the functioning of the circadian clock tends to weaken, reducing the physiological signals of day and night. This makes it harder for organs throughout the body to switch from day-appropriate to night-appropriate metabolism, blood pressure, hormone secretion, and more. The problems this change poses for the elderly, and for those who care for them, are discussed in Chapter 14.

So essentially all of us have built-in circadian cycles that are different from the standard twenty-four-hour day. And yet almost all of us follow twenty-four-hour cycles in our everyday lives. *Something* must be keeping our inner clock, and all the bodily functions that are affected by it, synchronized with external, solar time. And whatever that something is, however it works, it must be both accurate and ongoing. Otherwise, we would tend to drift further and further away from external time, like a clock that gains or loses a minute or two every day.

The most important cue our inner clock uses to keep itself in sync is the daily cycle of changing light. This should not be a surprise. After all, the sequence of day and night is at the root of our sense of time. In the biblical account of Creation, the first command was, "Let there be light!" And light is what Michel Siffre, deep in his cave, missed the most. In one unfortunate experiment of nature, blind people, who have lost all retinal function, have been found to free run just as if they were in a temporal isolation room. This happens solely because they cannot process the external signal of daylight to the eyes. At one point in the month they sleep soundly during the night, but a couple of weeks later, they are overwhelmed with sleepiness during the day. Some resist their daytime sleepiness and try to force themselves to get "normal" nighttime sleep, but they sleep badly and gain little rest. Others—especially if they are not constrained by work or family obligation—give in to their internal clocks and live their lives on a non-twenty-four-hour cycle.

Not all light is equal. In fact, the same kind and amount of light at different times of day has remarkably different effects on the circadian clock. Much of what we know about this is based on studies of lab animals that are exposed to periods of light at different points in their twenty-four-hour cycle. Unsurprisingly, it's the same for humans. The results show that light exposure has very different effects at different points in the cycle. In the morning, a period of bright light or a gradual dawn simulation moves the internal clock *earlier*. This makes it easier that evening to go to sleep naturally at an earlier time and to wake up earlier the next day spontaneously, without an alarm clock. A bright light pulse or gradual dusk simulation in the late evening, however, moves the internal clock *later*. This makes it easier to stay up late and to sleep later the next day.

In a Changing Light

In real life, of course, we are exposed to light throughout the day, and sometimes at night at levels that often vary wildly. Just stepping out of your home into an ordinary outdoor afternoon can multiply the amount of light you receive a hundredfold. The circadian clock does not ignore this real-life pattern. While dawn and dusk exposure have the greatest impact, the inner clock responds to light levels across day and night. To complicate the issue, different people experience vastly different daily patterns of light, and each of us from day to day is likely to have varied light exposure. If we measure the amount of light an average person gets moment to moment across the twenty-four-hour day, it would jump all over the place. As we experience more or less light from minute to minute, the circadian clock takes it all in and responds. But the nature of its response changes as the day progresses.

From early morning to early afternoon, whenever there is light input, the clock squeezes a little, or reduces the length of its daily cycle. For example, turning on the bedroom light might briefly change it from a 24.5-hour clock speed to a 24.48-hour clock speed. If you then walked into a darkened room, the clock would quickly stretch back to 24.5 hours. In other words, for most of the first part of the day, light exposure counteracts the slow inner cycle. But in the afternoon and evening, the circadian cycle stretches *longer* in the presence of light. If you squeezed throughout the morning, from a starting speed of 24.5 hours to 23.5 hours, and you stretched throughout the afternoon and evening by thirty minutes (0.5 hours), you would wind up in sync with local time, and achieve what is called *entrainment.* You're in balance with the outside world. This is what happens to most of us much of the time. It happens automatically, without our awareness or decision.

It is not just the direction of the clock's adjustments to a given level of light exposure that varies from morning to evening. The speed of the change varies as well. During the morning squeeze period, the power to squeeze gradually diminishes as we approach midday. An opposite pattern goes to work in the second half of the day and into the night. Gradually, light exposure gains more and more power to cause a stretch. This means that we are especially responsive to light exposure in the early morning and late evening hours. But the very same process persists even in the middle of the day—it just provides a weaker push in either direction.

Our internal clocks are continuously monitoring three main elements in the external environment. These are:

- The light level, moment by moment. (The more intense, the stronger the squeezing or stretching.)
- The net total of squeezes minus stretches in the first half of the day. (More squeezing to move toward a cycle length of twenty-four hours.)

- The net total of stretching minus squeezing in the second half of the day and most of the night. (Stretching needed to compensate for excessive squeezing that morning.)

The signals the clock receives are affected by many factors in our behavior as well. Are we keeping our eyes wide, squinting, or closing them a lot? Are we wearing sunglasses or blue-blockers? And as most of us know from experience, we can throw off the equilibrium by staying out late at night under bright lights. Not all hangover symptoms are just the aftermath of strong drink!

Faster and Slower Clocks

Even when two people have exactly the same amount of light exposure at the same times of day, the impact on their daily cycle may be different if they have internal clocks with different inherent cycle speeds. Someone with a relatively fast clock—one that cycles at, say, 24.1 hours—needs to make a net daily correction of only six minutes. Someone with a slower clock (cycling at 24.5 hours, for example) needs an adjustment of thirty minutes a day. And someone whose internal clock is very slow (say, 25.5 hours) needs to make a correction of ninety minutes every day to stay in sync with local time. This can present a real physiological challenge, given access only to the standard pattern of indoor and outdoor light exposure. People in that situation may adjust by falling asleep unusually late. In some cases, they may even go through periods in which sleep comes later day after day. It almost looks like they are free running around the clock, impervious to the time cues of day and night.

Wouldn't it have made more sense if living beings had developed inner clocks that accurately mirrored the twenty-four-hour solar day, without this need for constant adjustment? Not necessarily. A rough analogy might help explain why. Until the middle of the nineteenth century, local communities in the United States each had their own local time. Noon came when the sun was directly overhead, usually signaled by the church bell or factory whistle. Then came the railroads. If your train was scheduled to leave at 10:40, you couldn't rely on your local version of the time. You had to know that 10:40 came at exactly the same moment in your village, in the next town up the line, and in the city eighty miles away where the railroad dispatcher had his office. To be sure of that, you needed access to a clock that was both accurate *and* synchronized with other clocks in your area.

Therapy for Clocks

The railroads met this need by putting a big clock in every station. But even expensive clocks were prone to run a little bit fast or slow. And how were they to stay set to the same time as others along the line? The solution was to develop a clock that could be synchronized by a remote signal. These "Western Union" clocks, usually in handsome oak cases, still turn up in antique shops. When they were in use, an electrical pulse went out over the telegraph wires every hour, exactly on the hour. This pulse activated a magnetic mechanism that zeroed the minute and second hands of every clock in the network. This meant that even if the clocks weren't super-accurate (and super-expensive), they still kept exact *and* synchronized time.

Like a lot of analogies, this one is not really exact. Unlike a Western Union clock, we don't get a single correcting pulse every hour on the hour. Instead we get varying levels of light throughout our waking hours, which often differ a good deal from day to day. We may also be exposed to light at night while asleep, when it would be best if we weren't. The circadian clock responds to these light signals in a continuous fashion, as we've seen. At some points it is a little behind solar time, at some points a little ahead, but over the course of the day it usually does an amazing job of staying entrained to the twenty-four-hour cycle of the sun.

Not for everyone, however. For many people, the normal, often unpredictable day-to-day pattern of light exposure does not work on its own to keep their internal clock entrained. That is the situation that *light therapy* is designed to correct. Maybe your clock is running too slowly, your rhythms have weakened due to illness or age, or the lenses in your eyes have clouded to the point that you *need* extra input. You may be sleeping crazy hours that fall out of line with local time. You may not get enough natural light as the seasons turn, or as you dart from a dark home to a dark office. In such situations, the equilibrium of entrainment can easily go awry. Fortunately, we now have ways to put it right. These are the new techniques of chronotherapy, the focus of our book.

2.
The Pressure to Sleep

"Why do I have to go to sleep now?"

"Because it's bedtime, that's why!"

This little dialogue will sound familiar to anyone who has ever been in charge of children. Of course kids have to go to sleep at bedtime, and of course they sometimes resist that obligation. But on closer examination, both the child's question and the parent's answer raise other questions with difficult and even mysterious answers.

- *Why* do we go to sleep?
- Why do we *decide* to go to sleep at some particular (bed) time?
- Why do we sometimes go to sleep without wanting or intending to?

And most pressing for many, many people:

- Why do we sometimes—maybe even often—*fail* to fall asleep when we really want to?

Sleep is a feature of life that we share with every other mammal, every bird, and a great many reptiles and fish. Bears during the winter come out of their state of hibernation long enough to get some sleep. Whales sleep with one half of their brain at a time, while the other half takes care of breathing. Some birds, such as the swift and the arctic albatross, can apparently sleep while flying. It seems obvious that anything so universal, so insistent, must serve some absolutely essential function. But what? To conserve energy? To give the brain the quiet it needs to sort through the day's information, store what may be important, and dump the rest? To keep the person or animal in a protected place at times when there may be predators on the prowl? All these possibilities have been suggested, and all have met with objections. At this point, we can't really give a definitive answer to the question, "Why do we sleep?"

On the other hand, we know quite a lot about what happens if we *don't* sleep. And no matter what some high achievers and round-the-clock hedonists might think or wish, the consequences are very far from positive. Some are fairly trivial, such as having dark circles and bags under your bloodshot eyes. Others are much more serious. They include loss of alertness, difficulty concentrating, confused thinking, memory lapses, even hallucinations and delusions. Studies indicate that driving while sleep-deprived is as dangerous as driving drunk. And habitually getting too little sleep has been linked to an increased risk of developing diabetes and even of dying sooner.

Sleep is one of life's givens, then. Yet it poses serious problems for a great many people. Millions upon millions—about a third of the population—complain of sleep problems, or *insomnia*. Some have trouble getting to sleep at the start of the night. Some wake up in the middle of the night. Some wake up too early at the end of the night. And some combine any or all of these problems on any given night. In addition to the nighttime distress, they are less functional during the day. They feel more tired, less alert, and less positive. They may also develop eating and weight problems and other health issues. Insomnia usually occurs together with other health problems. Is their disturbed sleep the result of these health problems? A cause of them? Or—more likely, of course—are their sleep problems and other health problems reinforcing one another in a vicious circle?

There are many reasons people may develop a sleep problem. For example:

- They may have a nighttime breathing disorder that repeatedly wakes them up, such as obstructive sleep apnea.
- They may have taken something that affects their sleep, such as alcohol or a medication.
- They may be trying to cope with a circadian rhythm sleep disorder, the kind we are focusing on in this book.
- They may have some other medical condition associated with the insomnia. Among the possible suspects are mental disorders, such as depression and anxiety; neurological disorders, such as parkinsonism, epilepsy, and Alzheimer's; and physical disorders such as emphysema, asthma, and ulcers.

Given all the ways someone can develop insomnia, it is no surprise that a great many ways have been proposed to deal with it—and that so many of these do not work for a majority of sufferers.

Among the causes of insomnia, the circadian rhythm sleep disorders are most often overlooked or neglected. This happens even though they are common, can be easily diagnosed, and have effective, long-lasting treatments. Many people with sleep problems assume that their difficulties must be the result of some obvious external cause, such as work stress. It doesn't occur to them that the problem may lie with their inner clock, if only because they are not really aware of the clock in operation.

The result? They (and their doctors) are quick to resort to sleeping pills. But even if the pills work, they are only hiding the unhealthy lack of synchronization between their inner clock and the outer world. The problem will not go away *even if it looks superficially as if sleep has improved.* And as the world moves increasingly toward a 24/7 society, the sleep problems that result from a disconnect between inner and outer clocks will inevitably become still more common and pressing.

Time to Sleep

Why do we decide to go to sleep at some particular time? The answer that comes most readily to mind is some variation on the parent's answer: *Because it's bedtime, that's why!*

- If I [show up late to work/skip my nine-o'clock class] one more time, I'm in big trouble.
- If I'm sleepy and screw up that [conference/presentation/interview] tomorrow morning, I'm in big trouble.
- "If you don't turn off that [TV/computer/video game] and get to bed right now, you're in big trouble."

Or, more simply,

Uh-oh, look at the time. . . . Better get to bed.

In itself, sticking to a regular bedtime is obviously a good idea, but only if it fits well with the more basic chronobiological processes that help govern sleep. Basically, you have to let your internal clock tell you you're ready to get in bed and nod off. Artificial bedtimes you choose because of your work schedule, your spouse's preferences, your love of late-night TV—even your *conscience*!—can doom you to fitful sleep and poor daytime energy and alertness. The wrong bedtime can worsen any emotional problems you're trying to solve. Until you've changed your sleep pattern to fit your internal clock—or changed your internal clock, through chronotherapy, to fit your desired sleep pattern—a badly chosen bedtime can create sleep problems that range from mildly bothersome to deeply disturbing. This happens far too often. To make the needed changes, you first should understand how your inner clock functions, then use that knowledge to find the bedtime that is right for you.

The circadian clock is an exquisitely complex product of evolution. It guides when we sleep and when we stay awake by regulating a set of biological processes that include core body temperature and levels of the hormone *melatonin*, which is produced in the pineal gland. Both of these go through regular ups and downs across every twenty-four-hour day. During the afternoon, in the middle of what are normally our waking hours, body temperature is approaching its high point and melatonin level is at or near zero. As we move into the evening, body temperature starts to drop and melatonin level rises. Melatonin level peaks about midway into the sleep cycle, stays high until an hour or so before wake-up, and then quickly washes out of the system. Meanwhile, core body temperature hits bottom and starts to head back toward its daytime level.

When the level of melatonin starts to rise, this sends a message throughout the brain and body that circadian night is beginning, *no matter what time the clock on the wall might happen to say.* But if something interferes with this rise in melatonin, such as exposure to too-bright light during the evening, it disrupts the normal sleep cycle. As for core body temperature, the drop that happens during normal sleep hours is nature's equivalent of lowering the thermostat when the room isn't in use. The way the drop happens is that blood vessels in the skin become dilated. This allows more blood to flow through them. As a result, more body heat can radiate into the air. This process is the explanation for what seems like a paradox: As we approach the circadian night, we *feel* warmer, even though a thermometer would show that our core body temperature is actually falling.

The circadian clock controls other daily processes as well. A couple of hours before our normal wake-up time, just as body temperature is climbing toward its daytime levels, the clock signals the adrenal glands to release bursts of the hormone *cortisol.* By promoting the release of more blood sugar, cortisol helps get our biological systems ready to be up and about and to face the world afresh.

Sleep Pressure

From the moment we wake up, our bodies start building up a need to sleep again. The longer we are awake, the greater the need becomes. This need is known, logically enough, as *sleep pressure.* Once we do go to sleep, sleep pressure begins to recede. When the level of sleep pressure returns to its starting point, we wake up and the process begins all over again. In other words:

- While we are awake, sleep pressure accumulates.
- While we are asleep, sleep pressure dissipates.

If we are considering a "normal" day of waking up at 7 AM and going to sleep at 11 PM, we can see that sleep pressure accumulates fairly slowly, across the sixteen waking hours, and dissipates more quickly across the eight hours of sleep.

Days are not always normal, of course. Sometimes we stay up later than usual. Sometimes we wake up earlier or later. Occasionally we may even force ourselves to stay up all night. Whatever the circumstances, the same rules hold. Sleep pressure continues to build up as long as we are awake and starts to fade whenever we sleep. By the end of "a good night's sleep," sleep pressure will have gone down to zero. On the other hand, anytime we do not get enough sleep, it adds to a *sleep debt*. In effect, this means we start the day with a preexisting amount of sleep pressure.

One physiological process that may underlie sleep pressure is a buildup of a neurotransmitter called *adenosine*. This substance gradually accumulates during our waking hours and is linked to sleepiness. Significantly, caffeine targets the same neural receptors as adenosine, blocking its action. Presumably this is the reason that a cup of coffee helps us feel more alert and less sleepy. It would also explain why this boost doesn't last. The caffeine doesn't cut down on the amount of adenosine that has built up; it simply gives our brains a temporary break from reacting to the buildup.

Important as it is, sleep pressure is only one of the physiological processes involved in getting to sleep or not getting to sleep. The circadian cycle also has a powerful influence. It rises and falls spontaneously across the twenty-four-hour day, quite independently of sleep pressure. Each of these contributes to successful sleep, but what matters even more than either by itself is the way their two cycles relate to each other across time. Do they reinforce each other or come into conflict? The answer to this question is a key element in understanding both normal sleep and sleep problems.

Let's imagine that everything is working just as it ought to. During the day, sleep pressure gradually increases. However, because the inner clock signals that these are waking hours, we stay awake and functional. Toward the end of the evening, sleep pressure reaches a peak just as the inner clock is sending its nighttime signals. Barring any external interference, we fall asleep naturally and easily. In the morning, our system gets its wake-up signal from the inner clock, at a time when sleep pressure has already been reduced to its low point. We wake up bright and refreshed, ready to start the daily cycle anew.

But suppose we force ourselves to stay up all night. Maybe we have an important exam the next morning, or a report to draft that simply has to be on the boss's desk by 9 AM. Sleep pressure continues to build, but come morning, we find we can't fall asleep even if we have the chance and desperately want to. Why? Because the circadian clock has already issued its daily wake-up signal. At this point, that signal is powerful enough to override the ever-increasing sleep pressure. But at some point in the middle of the afternoon or early evening, sleep pressure increases to the point that it is able to take control. Circadian night is still hours away, but even so we may well crash. If that happens, we'll probably sleep longer than usual, right through the night, and wake up the next morning "on time" (at our usual circadian time, that is).

Siestas

The combined influence of sleep pressure and the circadian cycle do a good job of explaining why we go to sleep at night and wake up the next morning. But how do we explain siestas? These afternoon naps are a traditional feature of many cultures around the world. In the United States, a traditional midday meal on Sunday or Thanksgiving is counted a failure unless someone needs to stretch out on the couch afterward. Some stress-driven American corporations have decided it makes sense to set aside quiet, dim, lounger-equipped siesta rooms for employees. Big cities even have commercial "nap spas" with customer-adjustable lights, sound, and aromatherapy.

Once again, the best explanation we have at this time involves the relationship between sleep pressure and circadian rhythms. Sleep pressure starts to build when we wake up. Assuming we wake up in the early- to mid-morning, it will reach a fairly high level by early afternoon, roughly halfway through the waking day. Meanwhile, the morning burst of cortisol has worn off and the rise in core body temperature, another wakefulness signal, is only starting the gradual ascent toward its peak in the second half of the day. (It may even go through a small dip at this time of day.) The drive to sleep is strong enough to override the drive to stay awake and the result is that napping becomes a possibility. Food can be an added pro-nap factor. Eating a large or heavy meal at lunchtime causes an insulin response that leads to a temporary drop in blood glucose. This in turn promotes drowsiness. However, the primary reason that siestas are both attractive and possible is clearly circadian. Even someone who skips lunch can end up wanting a nap because of accumulated sleep pressure combined with low levels of wakefulness signals.

Is taking a siesta a good idea? Quite a lot of people have thought so, including Napoleon Bonaparte, Thomas Edison, Albert Einstein, and Winston Churchill. And lab studies back them up. After a nap, people are more alert and productive, less tired, and more positive in their mood. Their logical reasoning and decision-making skills improve, and if they have been learning something new or practicing a new skill, they retain it better after a nap. There is even research suggesting that regular napping reduces the risk of heart attack, stroke, diabetes, and obesity. What's not to like?

Still, some cautions are in order. It matters when we nap and how long we nap. Studies in which people were encouraged to take naps at different times of day confirm what most of us probably already suspect. For adults with a normal sleep-wake pattern, the best time to nap is in the early afternoon, about halfway between morning wake-up and evening bedtime. Napping later in the day, however, may have an unfortunate impact that evening. Naps deplete sleep pressure, and the later that happens, the less time there is for sleep pressure to build up again by bedtime. It is still making its way up toward the "ready-to sleep" level at bedtime, when your circadian rhythms of melatonin release and body temperature drop have already reached that point. Sleep pressure and the circadian cycle have gotten temporarily out of sync, and until they get back in harmony, you will have trouble falling asleep.

Researchers have also studied the effects of varying the length of naps. Surprisingly, it turns out that alertness and cognitive skills improve after as little as ten minutes of napping. That is enough time to relieve fatigue, too. Naps of twenty minutes or half an hour do not provide any greater benefits, and they are also more likely to set off a period of grogginess or *sleep inertia.* Those who take longer naps, of an hour or more, build up even more sleep inertia. There may be other gains from their deeper sleep, such as more creative problem solving, but it takes them even more time to return to full alertness and effectiveness.

Wake Maintenance

Imagine that you have to catch an early flight tomorrow morning. To get to the airport in time for all that security business, you figure you need to be up by 5 AM. Ordinarily you go to bed by midnight and get up at 7 AM, but you don't find the idea of getting by on five hours of sleep appealing. The solution seems obvious: Since you have to get up two hours earlier, you'll simply go to bed two hours earlier as well. At 10 PM you get into bed, turn out the lights . . . and spend what seems like forever lying there awake, fully alert. Why?

Of course you may be anxious about making your flight and about the trip itself. But even if you aren't, sleep will not come easily. The reason, once again, concerns the relationship between sleep pressure and the circadian cycle. In the early afternoon this relationship makes it more likely that you will doze off, but during the first part of the evening it makes it very *un*likely that you will be able to fall and stay asleep. Sleep researchers call this period the *wake maintenance zone* or, more picturesquely, the *Forbidden Zone.*

By the time evening arrives, sleep pressure is building up to high levels, especially if you did not get a nap during the afternoon. You might think this buildup is enough to let you fall asleep early. But now look at your circadian rhythms at this point. Your nighttime melatonin cycle hasn't even started. As for deep body temperature, it is still pretty high, having reached its daily peak only hours earlier. In the lab, when we measure deep body temperature continuously throughout the day and night, we find that natural sleep onset waits for the temperature cycle to go into its steep-decline mode, while skin temperature gets warmer as blood vessels in the skin dilate to help release heat. As melatonin enters the system it helps trigger this dilation, so these circadian rhythms work together to facilitate sleep onset. But until this happens, sleep pressure by itself is not enough to put you to sleep, even if you have been up a long time. *You have to wait until the two systems agree that sleep should start.*

Changes During Sleep

Not all sleep is equal. Scientists have known since the 1930s that during a normal night, we go through a regular sequence of changes in the characteristic patterns of our brain waves. The most important distinction is between REM (Rapid Eye Movement) sleep and non-REM sleep, but there are also different forms of non-REM sleep, which sleep researchers now refer to as N1, N2, and N3.

As we begin to fall asleep, we are still fairly alert. Brain activity starts to shift from small, fast *beta waves* to slower *alpha waves*. It is during this brief transition that we may experience vivid hallucinations and sudden, startling jerks. During N1, the real beginning of the sleep cycle, we are still rather close to being awake. This usually lasts for only five or ten minutes and is marked by very slow *theta waves*. With N2, which lasts for around twenty minutes, *sleep spindles*, bursts of rapid brain waves, make their appearance. With N3, slow, deep *delta waves* appear and gradually grow stronger. These mark the transition from light slumber to very deep sleep and are linked to the dissipation of built-up sleep pressure. This phase, often called *restorative sleep*, continues for about half an hour. At that point, the sequence typically moves on to REM sleep, passing through N2 again on the way.

REM sleep is named for the rapid eye movements that take place during this stage. It is also marked by faster breathing, by increased brain activity, and, most notably, by dreaming. At the same time, however, the voluntary muscles are taken out of the loop. We may have a vivid dream that we are running for a bus or away from a tiger, but luckily our legs stay still. As for why we have REM sleep, that is still not clear. One possibility is that it helps consolidate memories that have accumulated during the day. It may also encourage new associations and connections in the brain that lead to creative solutions to problems. Whatever insights we gain from future research on the topic, it is clear that REM sleep is doing *something* important.

Once the first episode of REM sleep is over, the cycle goes back to N2, then N3. Eventually it returns, via N2, to REM. In effect, then, we spend the night going back and forth between deep sleep and dreaming. We may have as many as four or five of these cycles during an ordinary night's sleep. Typically, the periods of N3 get shorter and those of REM get longer as we move toward a natural awakening. This is one reason alarm clocks can be so jarring; they may break into the vivid dreaming of the last, longest REM period.

How Much Sleep Do You Need?

Sleep is not an unwanted imposition or a luxury for the lazy. It is a biological and psychological necessity. But too often we do not get the sleep we need, and we suffer for it. It may happen only occasionally, with a late night out or an obligation that makes us get up much earlier than usual. We may regularly miss sleep on weekdays, and then try to make it up on weekends. Or we may be chronically short on sleep because of work schedules, family responsibilities, medical issues, or other problems. Three out of ten American adults get six hours or less of sleep per day, and even more report that they regularly get less than the recommended seven to nine hours. The consequences are wide-ranging and serious. They include a greater risk of obesity, diabetes, and heart problems; an increase in psychiatric conditions such as depression and substance abuse; and more motor vehicle accidents as a result of nodding off at the wheel.

How much sleep is enough? There is no magic number. Different people have different sleep needs. Some adults feel perfectly happy and fit even though they rarely sleep longer than six hours. Others feel they can't function without nine hours a night. And sleep needs change with age as well. If we are looking at averages, babies typically sleep most of the time, up to eighteen hours a day. By the time they reach school age, they generally sleep between ten and twelve hours. Teens should get up to ten hours a night, though not many do. As for adults, experts recommend around seven to nine hours. We should note that these are base figures. If you have been missing sleep, a *sleep debt* builds up. Even if you get an average night's sleep after several late nights, or a few in a row, that accumulated sleep debt may still leave you feeling sleepy and dull.

In deciding what your sleep goal is, you should try to avoid being too influenced by what you are told is "normal." The conventional eight hours of sleep is fine for many, too long for some, and too short for others. What matters is how you function during the day. If you wake up bleary-eyed, have to push yourself to get to work, find your mind wandering during the day, feel tired much of the time, want to collapse when you get home, and sleep in on your days off, you are probably accumulating a hefty sleep debt.

A good yardstick for judging whether you need more sleep is to look at weekdays versus weekends. If you sleep a couple of hours more on days off than on workdays, that indicates a problem. You are building up your sleep debt every working day, then trying—probably unsuccessfully—to pay it off at the end of the week. What you should strive for instead is to get the same amount of sleep every night. If you can manage to fall asleep just half an hour earlier, you will have gained three and a half hours of sleep across the week. You may have to skip the eleven-o'clock news or that late-night talk show, but you will get your reward in the form of improved energy, alertness, and mood throughout the week.

We should point out that more sleep is not always a good thing. Adults who habitually sleep ten hours or more a night may be dealing with *hypersomnia*, a condition that is closely linked to depression. If their depression is successfully treated, their sleep time goes down and their energy and alertness go up. Those with bipolar disorder also tend toward hypersomnia when they are depressed, which is most of the time, since manic episodes tend to come in relatively short bursts. But just before or during a switch to hypomania or mania, they suddenly start sleeping less but paradoxically feel in a better mood, with more energy and alertness.

What to Do

Do you think you may have a sleep problem that you should deal with? If so, the first step is to figure out the nature and size of the problem. Do you have trouble staying awake:

- While you're driving?
- During meals?
- Sitting at your desk at work?

- While doing housework?
- When together with your family?

A *yes* to any of these questions should raise a red flag.

Try keeping a sleep/meds log for several weeks: You can download a convenient log form (see *Resources for Follow-up*, p. 301). After you get up each day you give an estimate of when you fell asleep, if and when you were up during the night, and the time of final awakening, noting whether it was spontaneous or with an alarm clock. Along the timeline, you mark if and when you took medications or chronotherapy. Finally, you rate how you felt the day before in terms of mood and energy.

You will probably be surprised by what you discover. Our impressions about our own sleep often don't jibe with the facts. The log may help you spot connections you would not have noticed without it. ("Hmm . . .Why do I have so much trouble falling asleep on Thursdays?") It can also provide valuable background information if you end up going to a doctor for your sleep problem.

Take a close, systematic look at the *quality* of your sleep. It may look as if you are putting in enough hours, but if they are filled with restlessness, they won't do the job. Take the online sleep quality questionnaire (see *Resources for Follow-up*, p. 301). If your score is above 5, ask yourself if your problem might have a medical cause, such as feeling cold, pain, difficulty breathing, or leg twitching. If so, see your doctor. If the reason for the high score is not so obvious, it's time to consider the demands of your schedule, your mood state, insomnia without apparent cause—or a circadian rhythm disturbance.

Are you allowing yourself to wind down before bedtime? Or are you hooked on television, streaming movies, or the Internet? Turn down the lights in the evening, finish dinner at least three hours before bedtime, and if possible refocus on reading, family time, and calming music.

If you have to wake up early, try compensating with a slightly earlier daily bedtime. But don't forget about the wake maintenance zone. There will be a couple of hours before your normal bedtime when you can't fall asleep even if you're tired and sleep deprived.

If you take sleeping pills but still have trouble falling asleep for more than thirty minutes after going to bed, the pills are probably not working. Take the online *chronotype questionnaire* (see *Resources for Follow-up*, p. 301) to estimate when your circadian clock is signaling that it is time to fall asleep. If that's impractically late, you may be a candidate for chronotherapy. For guidance, take a look at chapters 7 and 8.

Even if you think your sleeping pills are working, consider the possibility that chronotherapy could give you a healthier result and help you rely less on drugs. Chapter 8 presents our analysis of the pros and cons of sleeping pills. If you've gotten into the habit of having a drink before bed, you should stop it! You may have to go through some difficult nights, but it is well worth the effort. Alcohol may help knock you out, but the sleep that results has poor physiological restorative quality and easily leads to daytime fatigue. Consider the chronotherapy alternative, to move your internal clock earlier and capitalize on your brain's natural sleep onset signal. Even those with serious alcohol problems can benefit from light therapy to help with the mood problems that often accompany their effort to dry out.

3. Owls, Larks, and Hummingbirds

Would you want your accountant to do your taxes in the afternoon, while struggling to stay awake until quitting time? Would your morning commute feel safer if the bus driver is still trying to emerge from a post-sleep fog? These are rhetorical questions, of course—would anyone really answer "Yes"? But the situations they describe are all too real. People differ from one another in a myriad of ways, and one important way is in their *chronotype.* The chronotype is a way of describing someone's habitual activity-rest cycle. When do you prefer to go to sleep, and when do you like to wake up?

One way to think about sleeping and waking is to look at physiological factors, such as core body temperature and melatonin levels, that help regulate the sleep/wake cycle. Chronotypes give us a different angle on sleeping and waking. They are not direct measures of internal physiology. Instead, they tell us about someone's behavioral preferences and abilities.

Suppose a friend suggests meeting at the gym and working out together. Would you rather do it in the morning or evening? If you have tomorrow off, when would you prefer to go to sleep tonight? If you had to stay up for much of the night, would you probably decide to get some sleep before, or after, or not at all?

Larks *vs.* Owls

Not surprisingly, the majority of adults have a chronotype that we can think of as "normal," or intermediate. These are the *hummingbirds*—a poetic way of describing those who stay smoothly in sync with the day/night cycle of the outside world. They generally wake up around 7 AM and go to sleep around 11 PM. During the day and evening, they manage to stay relatively alert and efficient, though they may feel a brief slump in the early afternoon. In our society, work schedules, store hours, mass transit schedules, and other social functions are generally organized around the habits of hummingbirds.

However, not everyone fits so neatly into this majority group, or into a world based on its needs and preferences.

- Some are *larks*. They wake up bright and early, before most people's alarms go off, and may start to fizzle out by early evening.
- Others are *owls*. They find waking up in time for work or school a constant struggle. Even when they do manage to get up and out, the fog doesn't really lift for a few more hours. Their day finally starts to get rolling in the afternoon, and if they try to go to bed at a "civilized" hour, they are likely to face insomnia.

We recently carried out an online survey of over five thousand adults ages eighteen to sixty and asked them about their preferred patterns of sleeping and waking. Overall, depending on where we drew the boundaries, we found that about 10 to 20 percent of adults would be classified as larks and a slightly smaller proportion as owls. The rest, some 60 to 80 percent, would be classified as hummingbirds, though many lean a bit in the direction of being owls or larks. When we look at men and women separately, women tend to be a little more larkish until around age fifty. Of course if we were to ask about a particular couple, both members may have very similar preferences. And if they don't—if one is much more of a lark and the other much more of an owl—this can quickly become a source of nightly tension.

While there are only slight gender differences in chronotype, the age differences are big ones. Children tend to be larkish, but at puberty that changes substantially. The level of melatonin production goes down and the inner clock shifts later. The greater freedom to set their own schedules and the increased pull of late-evening texting and Web contact with friends make a contribution, too. One by-product is an increase in evening light exposure, which further pushes the inner clock's sleep onset signal. Whatever the reasons, when teens are in a position to choose, most tend to prefer staying up late and sleeping late the next morning. Researchers refer to this as a *delayed sleep phase*, but we can simply think of it as getting decidedly owlish.

Once adolescents shift in the direction of owlishness, it can take years—even decades—for it to wear off. In our survey, we found a substantial difference in average chronotype score between the younger group (ages eighteen to thirty-nine) and the older group (ages forty to sixty). The two groups overlapped, but owls were much more likely to be in the younger group and larks in the older group.

Another way to describe the age-linked difference is to say that the inner clocks of the younger group are set to signal sleep time more than half an hour later than the inner clocks of the older group. And when we compare the youngest group, those in their twenties, with the oldest group, those in their fifties, the average difference grows to forty-eight minutes.

Did you ever wonder why the older guests at a party start to drift out the door and go home just when the younger guests are starting to get into the swing? It is not—or at least not entirely—a matter of different levels of energy or enthusiasm. What counts is that for the older group, the time on the clock is almost an hour closer to their physiological bedtime. They may be able to ignore, mask, or override the circadian night signal, but they cannot make it go away.

What Difference Does It Make?

Scholars and thinkers get up early in the morning and contemplate. Even those who desire wealth wake up early and get to work.

RIGVEDA (HINDU SCRIPTURES, ABOUT 1000 BC)

It is well to be up before daybreak, for such habits contribute to health, wealth, and wisdom.

ARISTOTLE (384–322 BC)

I never knew a man come to greatness or eminence who lay abed late in the morning.

JONATHAN SWIFT (1667–1745)

The sun has not caught me in bed in fifty years.

THOMAS JEFFERSON (1743–1826)

Early to bed and early to rise, Makes a man healthy, wealthy and wise.

BENJAMIN FRANKLIN (1706–1790)

No question about it, larks have enjoyed an excellent reputation for at least three thousand years. At the same time, owls have been generally looked down on as lazy and unmotivated. One of the few quotations we were able to find in praise of owlishness is from that well-known contrarian, Mark Twain (1835–1910): "Wisdom teaches us that none but birds should go out early, and that not even birds should do it unless they are out of worms."

This prejudice in favor of larks and against owls is deeply embedded in our culture. Read one of those admiring profiles of some successful entrepreneur. The first thing you learn is that he or she always gets up early enough to go for a five-mile run and put in some time with free weights. This is followed by showering, dressing, scanning the news, checking the European markets, and being the first to show up at the office. Oh, and did we mention that he or she has been using the morning commute to write an autobiography?

As far as we know, researchers have not yet tried to directly compare the job efficiency of larks versus owls. However, owls probably do not deserve their bad reputation. One study looked at reaction time, a measure of alertness, across the nine-to-five workday. The results showed that larks took steadily longer to respond to a signal, indicating that their alertness went down between morning and late afternoon. In contrast, the performance of owls and hummingbirds was slightly better at 5 PM than it had been at 9 AM. In another study, everyone's sense of alertness went up from morning to noon, but from that point into early evening, larks and hummingbirds became less alert, while owls continued to become still more alert.

So why have people who get up early gained such a positive reputation? We can imagine a lot of reasons, such as:

- Historically, in agricultural economies, some farmyard chores *had* to be done early. Dairy cows don't care what your chronotype is—they need to be milked first thing in the morning. Farm hands who did not get up soon enough to help were seen as not pulling their weight.
- Medieval monks thought of sleep as a sinful luxury and believed getting up to pray long before dawn was a sign of unusual devotion.
- Early risers reach their peak efficiency during the first part of the traditional nine-to-five workday. So the majority who are neither larks nor owls, including their bosses, are more likely to notice and admire their productivity.
- We generally think of the evening hours as a time for recreation—watching TV, going to a movie, hanging out with friends and family. Someone whose daily schedule provides more of these leisure hours obviously must be devoting less effort to more "productive" activities.

More generally, somebody who is asleep is clearly *not* up and working. And the fact that someone is asleep is more noticeable, public, and unusual during the day than during the evening. Two people may spend exactly the same amount of time asleep and awake, but the one who is always up and about at 6 AM will end up with a very different reputation from the one who usually gets up at 9 AM.

It seems that the Irish writer Robert Lynd was on to something when he said, "No human being believes that any other human being has a right to be in bed when he himself is up."

Where Did You Get That Chronotype?

The chances are very good that your sleeping and waking habits are more like those of your parents, brothers, and sisters, than like those of some stranger down the street. Researchers have found that chronotype has a strongly *heritable* component, maybe as much as 50 percent. For example, identical twins, who share the same genetic makeup, have more similar chronotypes than fraternal twins, who share only some of their genes. Genes are not the only factor that can make some characteristic heritable, however. For example, people often treat identical twins more similarly simply because they look alike. This, in turn, makes their social environments more alike. Still, the evidence is mounting that different versions of certain genes are linked to different chronotypes.

At the same time, the fact that someone's chronotype can change across the life span suggests that factors other than genetic background may also have a crucial impact. Do we know exactly what these factors are or how they work? No, not at this point, but we can suggest some of the usual suspects. The prenatal environment almost certainly plays a major role, as do a host of childhood experiences. To muddle the question further, what seems from the outside like the same experience or environmental factor may have a different impact on two children who have different genetic makeups.

Whatever the causes, every one of us ends up with a particular circadian clock that, as a general rule, does not quite coincide with the day/night cycle of the outer world. We can think of this as a basic *trait*, a biological and neurological quality that characterizes us throughout our lives. Within limits, however, it is adjustable. The passage of the seasons, a move to a different latitude—even something as seemingly trivial as the installation of a brighter streetlight outside the bedroom window—

can alter the *state* of our chronotype, or the way it expresses itself in our daily life.

Here is a case in point: We described earlier the tendency for teens to become more owlish. However, this does not mean that all teens become flat-out owls. Some remain larks or hummingbirds, if a little less larkish than they were before puberty. Now send them off to college. No parents around to announce, "Lights out!" and roommates and dorm mates who are themselves actually owlish. Add to that the general assumption that early bedtimes are for little kids and staying up late is a sign of adulthood. Late-night talk sessions, clubs that start to come alive after midnight, one more text exchange with someone back home or across campus, or one more round of that computer game, and before you know it these larks and hummingbirds are pulling down the shades on the bedroom window to shut out that pesky early morning light. Soon even someone with the *genetic* makeup of a lark can be sucked into having the *behavioral* profile of a total owl. As for those whose basic traits are owlish, they may find it impossible to get to sleep before dawn. Early morning classes? Forget it!

Know Your Own Chronotype

By this point, you probably think you have a pretty good idea about whether you are a lark, an owl, or a hummingbird. However, our ideas about ourselves sometimes mislead us. A famous example is the fact that in surveys, as many as 90 percent of people say that they are above-average drivers. If you want a more objective look at your personal chronotype, you can use the same measure that researchers use. Take the online chronotype questionnaire (see *Resources for Follow-up*, p. 301). This tool asks questions about your behavioral preferences and abilities. For example, when would you go to bed, and when would you get up, if the choice were entirely up to you? The questionnaire takes no more than five to ten minutes to complete, and as soon as you finish, you get personalized feedback. Your final score will fall between 16 and 86. Low scores—40 or below—indicate owlishness, and high scores—above 58—indicate larkishness. Most people, roughly 70 percent, fall into the hummingbird category. These intermediate scores indicate that the person's daily rhythm—sleep/wake cycle, alertness, energy—is in sync with local clock time.

The questions on our chronotype questionnaire focus on your behavioral preferences, but the score you get indicates more than simply the way you like to schedule your daily activities. In one study, we asked research participants to take the questionnaire, then to take samples of their saliva every thirty minutes during the four hours before their usual bedtime. They were also instructed to wear welder's goggles throughout the evening. The goggles kept the amount of light reaching their eyes to a very low level.

The reason we required goggles has to do with the hormone *melatonin.* Normally, at some point during the evening, a person's inner clock sends a nerve signal to the pineal gland, deep in the brain, to start producing melatonin for the rest of the night. The specific time at which this happens can vary from one person to another by as much as six hours or more. Once the pineal is activated, the hormone quickly shows up in the blood, and soon after in the saliva. A couple of hours later, you are ready to fall asleep. However, if you are exposed to bright light during the time melatonin production ordinarily starts, the pineal gland will stay shut down, as it is during the daytime hours. This robs the body of an important signal for normal sleep onset. The bright light also raises energy levels and triggers a delay in the setting of the inner clock. Dim light, such as that filtered through the welder's goggles, allows melatonin levels to follow their natural evening course toward bedtime and protects the inner clock from shifting later.

What did our study show? Based on their chronotype scores,

participants were divided into larks, owls, and hummingbirds. As we expected, the larks had melatonin levels that started rising as early as 7 PM. By around 9 PM, they were ready to go to sleep. For owls, however, melatonin did not show up in their saliva until much later, as late as midnight or 1 AM, and they were not ready for sleep until a couple of hours later still. So if you know your chronotype, based on our chronotype questionnaire, you also gain a good insight into the physiological functioning of your inner clock . . . and without the hassle of having to spit into a specimen jar every half hour!

Adjusting Your Pattern

Now that you know your chronotype, you can ask yourself whether you are content with it. If you are, Cheers! You are not alone. Many people are perfectly okay with their personal sleep/wake cycle. And not just the hummingbirds, whose pattern coincides with the norms and expectations of society. Larks and owls—even extreme owls—can get along fine if their circumstances allow them to live according to the urgings of their chronotype and inner clock.

Most of us have heard stories about brilliant Silicon Valley types who write computer code all night, sleep on the office couch, wake up around the time most people are going to bed, refuel with coffee and granola, and repeat the cycle. This is not just a myth or stereotype. A novelist friend has been waking at noon and going to bed around 4 AM for many years. She says she works best when it's dark outside. All her friends know there is no point trying to reach her before early afternoon. *Oh*—and her least favorite time of year, predictably, is the spring, when the days grow longer and the nights grow shorter.

But what if you are *not* happy with the daily pattern your chronotype mandates? What if it causes problems in your marriage or family or career? In that case, there are steps you can take to improve the situation. Owls, in particular, should consider these measures:

- Eat a protein-loaded breakfast soon after waking up, even if you are not hungry. Get your digestive rhythms in sync with your sleep/wake cycle, and you'll feel more alert and energetic in the first half of the day. One simple way is a protein shake, store-bought in single-serving containers, or prepared in the blender the night before, so it is ready to grab from the fridge.
- Stay away from coffee and other caffeinated drinks from midafternoon on.
- Move dinner earlier, so that you have three hours for digestion before going to sleep. It will make it easier to get to sleep and will also help align your circadian rhythms so that you can begin to shift from an owl toward a hummingbird chronotype.
- Emphasize carbs at dinner for their easy digestion and calming effect.
- Turn off your cell phone ringer and e-mail and text alerts at least one hour before bedtime. Since owls' friends are often owls themselves, tell them why you're doing this, so they don't feel neglected or ignored.
- Schedule exercise sessions for midday or before dinner, not in late evening. Evenings are for calming down, not for getting the energy going.
- Choose evening activities that don't get you excited or uncontrollably engrossed the way some TV programs and Internet activities are likely to do.
- Install dimmers on room lights and set low but comfortable levels in the evening. (If you are using high-efficiency, com-

pact fluorescent bulbs that can't be dimmed, switch over to the newer, dimmable types.)

- Dim your TV screen in the evening to a comfortable level that is no brighter than needed for comfortable viewing.
- Install the *f.lux* application on your home computer (see *Resources for Follow-up*, p. 301). This software cuts down on the activating effect of a bright white screen that can delay sleep onset by altering the kind of light your monitor gives off, depending on the time of day. If you're still at your workplace in the evenings, ask if you can install the app there, too. (Some, but not all, employers allow this.)
- Leave the shades up in the bedroom during the fall, winter, and spring, so you benefit from the natural dawn signal. (Depending on where you live, you should be careful in summer because dawn may be too early to do you any good, and might in fact turn you into *more* of an owl by delaying your internal clock.)

If you try these measures and they do not make enough of a difference, do not give up. In Chapter 7 we explain in detail how chronotherapy can act powerfully to bring your circadian system into line with the constraints of daily obligations and external time.

PART 2

Time, Light, and the Brain

4.
Getting Light into the Brain

It should not come as news that we need light to see, need our eyes to see it, and need a brain to process it into meaningful information. But exactly how does that all work? How does something *out there* become a mental image *in here*? In ancient times, the philosopher Plato thought that the eyes sent out rays. When the rays bounced off an object, they returned to the eyes to create an image of it. This may sound strange to our ears, but what Plato was describing is not that different from the way bats send out sound signals and guide their flight according to the echoes. The ancient medical authority Galen added that these rays from the brain must flow to the eyes through the optic nerve, which he imagined to be a hollow tube. The energy to illuminate an object to make it visible emanated from inside our body. This intriguing (but quite incorrect) assumption continued to influence ideas about vision for a thousand years or more, well into the Middle Ages.

Today our understanding of how we perceive light is very different. Though of course we look at and see visible objects, we don't project rays to illuminate them. Instead, the light enters our eyes, where specialized structures turn it into coded signals. These are passed along as nerve impulses to the parts of the brain that create vision. Separately, signals from the eyes also reach the internal clock.

There can be medical problems, however, that break this integrated system apart. Such cases are very sad but also informative. For example, a head wound might damage the occipital cortex at the back of the brain, which is basic to visual perception. The person becomes blind, but the circadian timing system continues to respond to light signals from the retina, preserving a normal sleep/wake rhythm. In a contrasting case, a tiny brain tumor grows at the base of the brain, invading the area of the inner clock. The body loses its ability to produce a circadian sleep/wake rhythm, and light signaling from the retina no longer syncs the clock to day and night. Sleep becomes erratic, in small, random bouts. However, visual perception remains normal: The patient can see the clock on the wall and know the time of day, but the information cannot be processed physiologically without an inner clock.

What can cases like these teach us?

For a start, we shouldn't think of the eyes and brain as different organs. The eyes emerge directly from brain tissue as the fetus develops during pregnancy, so we can more accurately think of them as *extensions* of the brain. The information we get from looking at the external world changes in form, from light energy to nerve impulses, and passes through a series of gates along the way. From the outside world, light passes through a transparent section of cornea, which covers the iris and lens, and enters the liquid chambers on its way to the cells in the retina, that respond to the light energy by producing nerve signals.

The various parts of the eye do not always work as well as they might, whether because of genetics, illness, or aging. When they don't work properly, the inner clock gets less light input, giving the light/dark cycle in the outside world less influence on our circadian rhythms. This in turn can lead to sleep disruption, energy loss, mood disturbance, or all of these—even including psychiatric disturbance.

If we start at the outside of the eye, we find that infection can make the cornea less transparent. This cuts down on the amount of light that reaches the pupil, the opening in the center of the colored iris. By the way, if you were born with blue or green eyes, you get a lot more light into your eyes than someone with darker irises. You may even be so sensitive to light that you have to squint when outdoors, depend on dark sunglasses, and wake up too early when dawn comes through your window.

As for the lenses, these go through a predictable set of changes across the life span. At birth, they may let dangerous ultraviolet frequencies get through, but by adolescence, these are filtered out. The lenses begin to get cloudy in middle age and by old age even become yellowish. This acts as a filter that severely reduces the amount of light from the blue part of the spectrum that can pass inside, though the person affected may not notice because the change comes on so gradually. Then there is cataract disease, in which a whitish mass develops in the lens that can eventually cause blindness if not corrected. One of us had cataract surgery some years back, and his lenses were replaced with artificial substitutes. He was flabbergasted, not just by the stark improvement in his vision, but even more by the way colors became so much more vivid. It was like washing years of grease and grime off the windows, leaving them sparkling clean.

The two liquid chambers of the eye serve mainly to provide the structural support needed to keep the eyeball round. They, too, have to remain clear if the system is to transmit light properly. And they, too, can easily cloud up in middle age, causing decreased light transmission and an experience of visual glare as photons bounce around after hitting the tiny particles that form the clouding.

Finally, we come to the retina, the dense field of visual processing neurons at the back of the eye. Behind the processing equipment lie the photoreceptors, which absorb light energy, in the form of photons, and change it into neural signals that can be transmitted to the brain. One set of photoreceptors—the rods—becomes sensitive during darkness and enables us to sense the dimmest light signals. Another set—the cones—becomes sensitive during the daylight, and is specialized to see color and detail when light intensity is high.

As darkness ends and daylight approaches, the pigment in the rods gradually bleaches from light exposure, so the rods become far less active until dusk returns—or until we step back into a dark environment. If you go to a movie matinee and arrive after the film has started and the theater is dark, it takes several minutes before your eyes can tell you if that seat you want is already occupied. This is because the rods are not yet operating efficiently. Then, after the film is over, when you walk out into the afternoon light, you will probably squint and shade your eyes as your pupils contract protectively, while your rods bleach out in response to the new level of brightness.

It may seem like an odd anatomical arrangement, but the rods and cones are actually at the back of the retina, behind a forest of other nerve cells. Once light reaches them, they send their neural message forward, toward the frontmost layer of ganglia (singular, *ganglion*), which consolidates information from the processing cells that lie in between. There the message about color, shape, intensity, spatial arrangement, and movement is finalized for entry into the brain. The ganglion cells then project backward to the exit pathway from the retina, the optic nerves. The optic nerves transmit the light information to the brain, both to form visual images and to signal the time of day to the inner clock.

A Clock in the Eye

The more we learn about the circadian timing system, the more nuances and intricate mechanisms we seem to find. In our research, we discovered that the retina itself contains clockwork that affects the amount of light needed for a signal to be passed along to the brain. This specialized internal clock is self-contained in the eye and operates independently of the master clock in the brain.

We learned this by training lab rats to "report" (by pressing a switch) whenever they saw a dim flash of light. They lived in a darkened testing chamber and received a reward when they reported correctly. The darkened chamber allowed their rods to remain as sensitive as possible. However, the level of flash they could detect did not stay the same at all times. Their light sensitivity moved higher at night and lower during the day. This cycle did not exactly follow the twenty-four-hour solar day, by the way—but rather the *circadian* day, which was slightly different in length from one animal to another.

We also learned that the rods' cycle of visual sensitivity was not under the control of the clock in the brain. In a delicate operation, we surgically disabled the brain clock. The circadian cycle of activity disappeared immediately, but the cycle of visual sensitivity continued. The eyes had kept the ability to cycle even when this was lost to the rest of the body.

What was this clock-in-the-eye doing? We guessed it was modulating the sensitivity of the rods, since these are the visual cells that process low light in darkness. In the next experiment, we put animals in a light/dark cycle so they would show a normal rhythm of heightened rod sensitivity during the night and low sensitivity during the day. Under the microscope, we saw that the rods grew longer during the night and then were suddenly shorter about two hours after the lights turned on. Packets of the sensitive visual pigment *rhodopsin* were lopped off in the morning and then gradually replenished over the next twenty-four hours. In effect, the tips of the rods wore out even with low light exposure at night, and new rhodopsin had to be produced for them to be sensitive during the next cycle. This circadian process is called *photoreceptor renewal*, and we humans have it, too.

We then tried leaving the animals in continuous darkness. The rods still shed their tips the next morning, even though the lights hadn't come on. So the shortening of the rods was not caused directly by light exposure. This looked like internal clockwork in action—but where? The brain's inner clock, or the independent inner clock in the retina of the eyes? When we surgically disabled the brain clock, the results didn't change. Under a daily light/dark cycle, the rods shed their tips about two hours after morning light turned on, and under continuous darkness they continued to shed their tips once a day. Clearly it was the circadian clock in the eyes that was in charge.

Why did it take two hours after the lights came on for the rods to shed their tips? We worked with colleagues in Switzerland to answer this question. The key was to recall that in nature, morning light does not come on suddenly with the flip of a switch. But in our lab—as in most labs—it did. When we tried giving the animals gradual dawn simulation instead, the rods responded by shedding their outer tips *hours sooner*, at the faintest beginnings of morning light. This probably happens with us as well. Yes, even if we flip on the lights the moment the alarm goes off or sit in the dark all day, the eye's clock will make sure that receptor renewal eventually takes place. But if we sleep with the shades up (or use a dawn simulator in the bedroom, as we describe in Chapter 7), the renewal cycle gets under way much, much sooner. The day begins as nature intended it to: smoothly. The inner clock appreciates the difference in input from the eyes, and responds by resetting. Waking up becomes easier—and earlier! That is the essence of chronotherapy.

A New Response to Light

Scientists who use lab animals to study vision have known for a long time that some blind mice continued to show a daily light/dark cycle in their physical activity. Some of these animals were born without rods and cones. Others were affected by a condition similar to retinitis pigmentosa—a slow degenerative disease—in humans. In either case, the animals had essentially no visual receptors in the retina. How, then, could their inner clocks still be responding to light? The best guess was that they were not totally blind, that a few remaining vision cells had somehow escaped detection under the microscope.

To check out this possibility, researchers deliberately bred genetic strains of mice that lacked *all* rods and cones, and as a result were totally blind from birth. Even so, their inner clocks were still in place and still functioned. When kept in continuous darkness, their rest/activity cycle cycled at less than twenty-four hours, just as it does for sighted mice. When exposed to fifteen minutes of light near the middle of their nocturnal activity period, their inner clocks reset by an hour and a half—just as they would have for sighted mice. And when they were placed on a twenty-four-hour cycle of artificial lab lighting, they became active at night and quiet during the day, just as any healthy nocturnal animal would do. This bears repeating: *These mice were totally blind*—they did not have *any* visual receptor cells in their retinas. Even so, their inner clocks were still responding normally to light and dark. How could this be possible?

The breakthrough came in the 1990s, when researchers isolated an additional light-sensitive neurochemical in the retina, one unknown to science until that point. At first, many scientists were unconvinced. After all, the eye had been intensively studied for two hundred years or more. But the evidence continued to pile up. How had everyone missed something so important, for so long? The answer may be that they had been paying most attention to the visual photoreceptor cells—the rods and cones. But this new substance, given the name *melanopsin* (not to be confused with melatonin), was not present in either rods or cones. Instead, it was located inside a small proportion of the ganglion cells that make up the front layer of the retina. The retina has roughly one and a half million ganglion cells, but only 1 to 2 percent of them, about fifteen to thirty thousand, contain melanopsin. Given that the melanopsin is relatively scarce *and* sits in a totally unexpected place, perhaps we should not be too surprised that its existence was overlooked until so recently.

The discovery of melanopsin provided two new insights into the way the inner clock works. First, some ganglion cells in the retina are able to respond to light even if they do not receive signals from the rod and cone visual receptors. And second, this small number of ganglion cells enable the inner clock to stay in sync with external light/dark cycles, even when the rods and cones are totally missing, as in those genetically altered mice. When this work was extended to humans, the same effects were found. We now have a cautionary note for ophthalmic surgeons when they consider removing the eyes of patients who are blind from birth or as a result of retinal disease. Even though their rods and cones do not work at all, they may still have ganglion cells containing melanopsin that keep their biological clock synchronized. As we have said, some blind people have free-running circadian rhythms that have become untied to the external day/night cycle. This suggests that their melanopsin-containing ganglion cells are missing or nonfunctional. But if the patient shows any circadian rhythm response to light, the eyes should be left in place whenever possible.

A further insight from the animal lab was soon to come: Melanopsin turns out to be specifically sensitive to blue light (and also to white light, which contains a major blue component). But there is an additional twist to the story. The research on mice shows that the light-sensitive, melanopsin-containing ganglion cells can operate independently of the visual rod and cone photoreceptors, and provide synchronizing output to the circadian clock. In normal operation, though, these ganglion cells also respond to rod and cone stimulation from light across the entire spectrum, not just to blue. In other words, visual stimulation that creates activity in the rods and cones can also modulate the activity of the circadian ganglion cells. They can increase or decrease their sensitivity, probably depending on when during the day the light exposure happens. Exactly *how* this works is still unclear, and is the focus of current research. We are reminded of a set of nesting puzzle boxes: Figure out how to open one, and your reward is the chance to try your hand at opening the next one inside.

From the Eye to the Brain

Nerve impulses from the retina leave the retina by way of the optic nerve. This is like a cable made up of fibers from all the retinal ganglion cells. Of course, there are two optic nerves, one from each eye, and shortly after entering the brain, they meet and partially cross at the *optic chiasm.* (The term comes from the Greek word for the letter X, which is what the meeting of the two nerves looked like to early anatomists.) From there they continue on their way toward the visual cortex, the part of the brain most responsible for visual perception.

Not all the nerve fibers from the eye make the complete journey to the visual cortex, however. A smaller bundle branches off and goes toward the hypothalamus. This brain region sits right above the optic chiasm and is responsible for monitoring and controlling many essential processes, such as hunger, thirst, and body temperature. In the early 1970s, researchers at the University of Chicago managed to trace these specialized optic nerves directly from the ganglion cells of the retina to a tiny area at the base of the hypothalamus, no bigger than a grain of rice. Because it is located directly above the optic chiasm, this area is known as the *suprachiasmatic nucleus*, or SCN.

At about the same time that the Chicago group identified the link between the retina and the SCN, researchers at the University of

California, Berkeley, found that when the SCN was disabled in rats, the daily activity cycle vanished. This was the first clear evidence that the SCN plays an essential role in controlling circadian rhythms. In the decades since, more and more details have emerged to prove that the SCN is in fact the master clock, not just in rats, but in mammals generally. That includes us humans.

There are other biological clocks in the body, including one in the liver that helps control the digestive process and is sensitive to when you eat, but the most important for sleep, mood, hormone, and temperature regulation is that little group of about twenty thousand neurons, sensitive to retinal light input, called the SCN.

When we consider the importance of the SCN for circadian timing, we gain a new appreciation for the role of the recently discovered photochemical melanopsin. The ganglion cells that contain melanopsin are also the ones whose nerve fibers go straight from the retina to the SCN. This means that critical light signals can reach the circadian control center in the brain even in an animal or human that is totally blind.

Now we can better understand the reasons that the two situations we described at the beginning of the chapter ended so differently. The patient with brain damage to his occipital cortex continued to live according to a normal circadian rhythm because his retinal ganglion cells and SCN were still operating. On the other hand, the patient whose brain tumor pressed against the SCN lost daily rhythmicity even though he still could see.

These recent discoveries about the eye and the brain teach a major lesson to all of us—beyond the pathology experienced by patients like these. If you trip over a lamp cord and fall, you can feel the pain and see the bruises develop. You have no trouble drawing a clear connection between the fall and its painful consequences. The result is that you pay more attention after that, and you move the lamp cord to keep from tripping on it again. In short, when we run into a problem in our everyday physical world, we have no trouble linking cause and effect, then looking for ways to solve the problem.

It is not so straightforward with our inner circadian rhythms. They do not jump up and down and bark at us. We may not even notice them at all. So when they affect our mood, or our eagerness to go do something, or our ability to get to sleep, we are less likely to trace these effects back to their cause and decide to do something about it. Yet these rhythmic, light-sensitive cycles have a profound impact on us from moment to moment and day to day.

This is particularly true today, because we now live in a world where artificial light is everywhere. Whenever we expose ourselves to light, it signals our inner clock—even when we are not consciously aware of the exposure. It is a bit like the nineteenth-century discovery that many diseases are caused by microbes. We still can't see them without special equipment, but knowing about them has let us counteract some of their worst effects. Once we begin to understand how light from the outside world signals the circadian timing system and how resetting the clock can affect our well-being, we can begin to take control of the process for our benefit.

5. Getting Light Wrong

The virtues of light and the dangers of darkness have inspired poets and religious leaders since the dawn of history. Around 1500 BC, the prophet Zoroaster told his followers that Ahura Mazda, who existed in light, created everything pure and good, while Angra Mainyu, who dwelled in darkness, was the source of everything evil. In the Bible, the third and fourth verses of the first book, Genesis, proclaim: "And God said, Let there be light: and there was light. And God saw the light, and it was good: and God divided the light from the darkness." In the views of some evolutionary thinkers, our brains are constructed in ways that make us more likely to value light and fear darkness. The explanation they offer? During daylight our very distant ancestors could hunt and eat, but when the dark of night came they were more likely to be hunted and eaten themselves.

Whatever the situation in prehistoric times, and however much or little its carryover may still affect us today, in our contemporary world we experience light in many ways that lead to problems.

- We may not get enough light even when it is available outdoors.
- We may get light, but at the wrong time of day (or at night)—especially artificial indoor light.
- Unlike our distant ancestors in the tropical regions, we may be enduring very long dark winter nights and all-too-brief winter days.
- We may have circadian clocks that put our physiology at odds with external time.

Life by Day

There are now some seven billion people in the world, and a huge proportion live in urban areas that sentence them to a largely indoor existence. When populations become dense, space becomes tight. Even in ancient Rome, the solution to that problem was the multi-floor apartment house. The Romans had no elevators, so their tenements rose no higher than seven or eight stories, but today there are few limits. We live surrounded by walls, floors, and ceilings. Windows? Often covered by blinds and curtains, in the hope of gaining a bit of privacy from those all around. Or blocked by the adjoining building, the sky a small, distant patch of blue. And even when a window faces an open, unshaded view, a light meter held near it in the room will register much lower than it would just an arm's length outside.

"But I keep the lights on, even during the daytime." Okay—but even without considering your electric bill and enlarged carbon footprint, that is far from doing enough for you. The best-lit architect's studio or hospital surgery suite still registers in the upper twilight range. Because we have become accustomed to living all day long in twilight, it feels perfectly bright and comfortable, but in physical terms it is still dim. The situation extends to the workplace and to the deliberately shaded cars, trains, and buses that carry us from home to work and back again. As for those in institutions—hospitals, retirement homes, prisons—they are often "protected" from direct outdoor light twenty-four hours a day.

Our forebears, the hunters and gatherers and farmers, routinely experienced daylight intensities much greater than 1,000 lux. Today, the chances that an average urban (or suburban) dweller gets an unblocked dose of outdoor light of that intensity for more than a few minutes at a time are very small.

And by Night

Imagine that you are standing in an open field, far from the skyglow of cities, on a clear moonless night. The array of stars that fills the sky seems amazingly bright, especially if you have given your eyes time to dark-adapt. In fact, however, the light that is reaching you amounts to no more than around 1/10,000th of a lux. To our inner clock, that counts as *darkness,* and darkness is exactly what our inner clock expects and needs once night falls.

But today? We are assaulted by artificial light all evening and all through the night. Yes, artificial indoor light falls in the twilight range, far below daylight. But to our nervous systems, twilight is *not* the same as nighttime. A dimly lit living room is still bright enough to signal the circadian clock that night has not yet arrived. The same is true for light from the television screen or computer monitor.

Eventually, some combination of sleep pressure, circadian signals, and long-set habits—hopefully not including sleeping pills—sends us to bed. But there, too, light pollution assails our system. Even a night-light, improperly designed or badly placed, can throw off our circadian clockwork and disturb sleep timing and quality. Light leaks from outside also penetrate the bedroom. There are streetlights, a neighbor's motion-detecting yard light, the headlights of a passing car sweeping across the ceiling. . . . These can turn the night into a mishmash of dark and light signals that leave the inner clock in a state of deep confusion. Are you sleeping under starlight? Not likely.

Long Winter Nights

Our species evolved in regions near the equator and spent most of its two-hundred-thousand-year history there. These distant ancestors experienced about twelve hours of daylight, three hours of twilight, and nine hours of darkness all the year around. But a time came when much of humanity migrated toward the poles—or better said, at least halfway toward the poles. Obviously, they were able to survive this move, or we would not be here. But there was a cost, especially for those whose genetic makeup was more strongly locked into the twelve/three/nine equatorial day. In winter, as the nights grow longer—up to eighteen hours or longer in northern regions—our more vulnerable brothers and sisters become listless, feel intense sleep pressure, start gaining weight, and become markedly less productive and engaged, compared to the way they are in summer. This is the very picture of depression, but *seasonal* depression.

If you trace a line on the globe at 38° North, it passes near Ankara, Turkey; Beijing, China; Lisbon, Portugal; and Washington, D.C. That may not sound very far north at all, but to our physiological systems, it is. Among those who live north of that line, about 5 percent—one in every twenty people—suffer significantly from seasonal mood problems. If we go an equivalent distance south of the equator, the population density is much lower, but both the problem of longer nights and the frequency of seasonal mood problems are similar, only displaced by six months.

Slower or Faster Clocks

Along with vulnerability to long winter nights, there is a second genetic factor that confounds the healthy daylight exposure pattern for many people. This is the inherent speed of our circadian clocks. If we isolate ourselves from the external day/night cycle, our physiology would put us on a daily cycle of sleep, mood, and energy that is longer than twenty-four hours. That is what happened to Michel Siffre during his long cavern stints, which we looked at in Chapter 1.

The cycle length of the clock in the brain is set by a person's genetic makeup, which is particular to the individual. One person's inner clock may run at twenty-four hours and fifteen minutes, while another's is closer to twenty-five hours. This may not sound like a lot, but it makes a large difference. It is not hard for the inner clock to make a fifteen-minute daily adjustment in response to morning light exposure. Such an adjustment would bring the person with the shorter cycle into sync with twenty-four-hour solar time. However, it is much harder for morning light to subtract a full hour from the internal cycle, so the person with a longer cycle faces a much greater physiological challenge. Internal clock pressure can become so strong that it pushes the daily rhythm later, despite the timing of outdoor sunrise, until it is impossible to fall asleep and wake up according to local time.

Paying a Price

There are many ways we subject ourselves to inappropriate lighting, including contemporary urban light deprivation, nighttime light pollution, the change in seasons as we move farther from the equator, and the pressure to reconcile a slow-speed internal clock with external time. When we do so, we may suffer major negative consequences. These include:

- Gloom and despair lasting weeks or months
- Excessive sleep
- All-day sluggishness
- Irresistible food cravings
- Inability to get work done
- Self-imposed isolation
- A sense of pervasive anxiety

The first light-deprivation syndrome that research clinicians identified and studied was winter depression, soon to be better known as seasonal affective disorder, or SAD. Remarkably, it was a patient, Herb Kern, who spurred on this discovery and the work that followed. Kern, a chemical engineer, had kept a diary of his seasonal ups and downs for years. When he learned that bright light could stop melatonin production by the pineal gland, it occurred to him that this might help explain his depressive slumps in fall and winter.

Kern took his new insight and the extensive record of his moods to the National Institute of Mental Health (NIMH). There, an intrepid team of researchers, Thomas Wehr, Norman Rosenthal, and Alfred Lewy, tried "turning winter into summer" with a light box—just a standard fluorescent ceiling unit like we have in office ceilings, but set on a table close to where he would sit. It worked! Kern quickly snapped out of a deep depression. This was lucky for him, because he could not tolerate antidepressant drugs or lithium, the "mood stabilizer." He later became a coauthor of the first two medical journal articles on SAD, in acknowledgment of his unusual role as a patient who directly and actively helped spur medical progress.

Explaining the Blues

People have probably known about winter depression for as long as members of our species have lived in regions with marked seasonal change. And for just as long, people have been developing plausible explanations for this distressing state of mind. We are still at it. If you ask your friends why somebody might feel blue during the winter, you will probably get quite an assortment of answers.

- There's more tension at work in winter.
- There's all that stress over school papers and finals.
- It's been too long since last summer's vacation.
- It's too long to wait until next summer's vacation.
- It's too cold.
- Getting home from work or school when it's already dark out is a drag.
- Going to visit family over the holidays is too stressful.
- Not having family to visit over the holidays is too sad.

Then there is Ebenezer Scrooge's explanation, in Dickens's *A Christmas Carol*, that the holidays are a season for finding yourself a year older and not a penny richer!

All of these explanations may have the ring of truth for one person or another. But what all of them overlook is a radical possibility. What if these social and psychological problems, real as they are, have been amplified or even created by the gap between what our biological systems expect from the outer world and what they are actually getting? If that is so, the best way to address the problems would be to try to eliminate that gap.

Eleanor's Story—Part I

Eleanor T., fifty-four, is a psychologist who lives and works in the Boston area. Her easy smile and pleasant manner would not lead you to think that she has been dealing with seasonal mood problems for most of her adult life.

"I first got winter blues in college," she recalls. "That's when it hit, but I didn't know what it was. Intuitively, I knew I was better in springtime. As soon as the sun came out, I would lie outdoors in the Public Garden. I felt better with the sunshine, in spring, summer, and the fall. You know, I worked outside every summer in college, getting exercise and light, and it felt great."

Eleanor tried to overcome her mood problems on her own. "Initially, I found that walks in winter and aerobic exercise helped, as did moving to a sunnier apartment," she says. "I have a lot of sun in the bedroom, and that's helped a lot. But I still have ups and downs. I have mood swings, sometimes they're worse. When they are, there have been times I have worried that I would develop bipolar illness. My family history includes bipolar.

"I would say that I have mixed mood states," she adds. "Sometimes I am very depressed, and sometimes I'm unable to sleep or sit still. There were times I felt so depressed it was hard for me to pick clothes up off the floor. I would have to push myself to file or do laundry, things that take no energy or thought when I feel okay. Another example: Something that reliably makes me feel good is being outdoors and seeing something beautiful. When I was really depressed, though, there were times when I would go outdoors and I felt *nothing*."

Eleanor's friends did not necessarily realize that she was trying to cope with depression. "Even if I'm feeling depressed, I'm able to pull things together at work, or in recreational social situations. So the things I had most difficulty with are the things I have to do alone. I see it with my own patients all the time. They say they're depressed, but they look fine—you can't tell. Some of us feel better when we're talking with people, so the situations where I felt worse were doing things alone—like taxes. You're doing it alone, and it's boring."

Sleep was a problem as well. "I tend to avoid pills if I can't sleep, so there were times I had trouble sleeping through, or sleeping soundly. What comes back to my mind is a conference on Cape Cod in July. The schedule was meetings from nine to noon, and so I was spending a lot of time outdoors, and the days were long. That week I slept wonderfully and I felt great."

SAD, Depression, and Bipolar Disorder

Since the breakthrough by Kern and the NIMH team, research on SAD has shown that its symptoms are much like those of major depressive episodes that can occur at any point in the year and last for weeks, months, or unremittingly for years. What distinguishes SAD is its timing, with its clear link to wintertime light deprivation.

A clinical depression usually involves a plummeting mood of deep sadness, tearfulness, and a sense of misery. Some experience it differently, becoming "emotionless," unable to find the motivation to achieve at work, see friends, or participate in family activities. *Blah*, rather than sad. Most people experience it both ways, blah and sad. When this kind of experience is absent in summer but returns almost every winter, we have SAD.

The SAD experience in summer comes in two varieties—calm and peace with the world ("euthymia" in medical jargon) or hyperenergized ("hypomania" or even mania). People who experience the latter have seasonal *bipolar* disorder, which at its extreme can be very disruptive. About 20 percent of people with SAD show this pattern. The symptoms come in the spring, just when the winter depression lifts, and often taper down after several weeks. Even people with *nonseasonal* bipolar disorder—with depressions that can occur at any time of year—frequently switch high in spring, which indicates that they, too, are sensitive to the post-winter increase in outdoor light.

Roberta's Story

Roberta R., fifty-nine, is single and lives near Cleveland. She is retired and depends on Social Security/Disability. She has been hospitalized four times for major depression and bipolar disorder.

Her first hospitalization came when she was living in Illinois and working for a freight company. "My nephew was living with me," she recalls. "I left work one day and when I got home, I told him I didn't remember how I got there. I said, 'You have to take me to the hospital.' The hospital told me I couldn't work anymore. The company where I worked filed Social Security for me, and I got it, but it was a huge cut in my income. I had to file for bankruptcy."

Roberta moved back to Cleveland eight years ago. "I've been living with my father. He was taking care of me for a while, and now he's ninety-two years old, and I'm taking care of him. I can't afford to get sick now. My father had trouble with my depression. He thought, 'Oh, just snap yourself out of it.' But my sister and her son and grandson—ALL bipolar. So she understood some of what it's like for me."

Roberta's psychiatrist put her on antidepressants. "I was on the meds for a really long time," she explains. "But I would still have bad depressions. Then I joined a bipolar support group. Several people talked about SAD. That made a big impression on me. I knew I'd get worse during the winter. I was depressed at other times of the year, but all of my hospital confinements were in the fall and winter."

When Roberta heard about a study that involved light therapy, she applied to be part of it. "I went in and talked to them. They gave me a light unit and I went from five minutes to about fifteen minutes, and noticed I was feeling better. Eventually I went up to about an hour, and it broke the depression. The lights made me feel better. I always knew that I was better when there was a lot of light."

But the study ended. "The insurance wouldn't pay for the light box, and I couldn't afford it," Roberta says. "I had to give that light box back. I spent a lot of time outside—that helped. But it's been such icky weather—rainy summers—and in the winter I'm stuck in the house. And it's been wearing on me. Last month, everything got worse. I was more depressed and I didn't want to go up on the meds. I was starting to slide. I knew winter was coming. I was possibly going to have to go to the hospital again, and I didn't want to."

Then Dr. Dorothy Sit, Roberta's former psychiatrist who now conducts chronotherapy at the University of Pittsburgh, called to ask if she was willing to be interviewed for this book. "She asked me about the lights," says Roberta, "and I said I couldn't afford it. She told me the insurance companies are paying for it now. I checked, and they said they would give me one. So I started the lights on Wednesday, five days ago, for five minutes, and it makes me feel better. On the seventh day I increased to fifteen minutes. Just getting the light makes me feel better."

Roberta laughs. "They showed me pictures of what I looked like before and after that first study. It was BAD. I looked terrible.

It was like day and night. I knew I felt terrible then, but I didn't realize I looked as terrible as I did. When you're really depressed, you kind of block things out."

Recently Roberta redecorated her bedroom. "It's much brighter," she says proudly. "I got a comforter and sheets and pillows in nice summery blues and yellows. I guess I'm trying to get the summer back! Every time I come into the bedroom, I feel happy."

The "high" side of bipolar disorder leads past euphoria to a distinct downside that requires intervention. If people around you start to comment or complain, it's an important signal that the high has reached a clinical level that requires professional attention—*"But I've never felt better in my life!"*

Consider these hallmark symptoms:

- Your mood changes, so you feel either unusually good or optimistic, or else irritable and argumentative.
- You feel little need for sleep, and may get two or more hours less sleep than usual nightly.
- You're fidgety and distractible, and your thoughts race so quickly your speech can barely keep up with them.
- You are sure you're being creative and insightful, even if you're not.
- You're impulsive with people and activities, and strike up conversations with strangers and long-lost acquaintances, volunteer for way too many assignments at work, spend a fortune shopping, or have flings you would normally never dream of.

No individual with bipolar disorder will show all these symptoms, but that's what to look out for, in yourself and in those around you. These unhealthy highs can be treated by a variety of medications. Often, patients take a mood-stabilizing drug such as lithium—not only to calm an episode, but to protect against recurrences that are otherwise very likely.

Here is where the light/dark story comes in again. We've learned recently that staying in a dark room for up to fourteen hours can break a manic episode, just as fast as the strong antipsychotic drugs often used in emergencies. Patients also benefit by wearing circadian blue-blockers all day, glasses that cut down on the amount of light in the blue part of the spectrum. As for the connection to mood-stabilizing lithium treatment? Swiss researchers Anna Wirz-Justice and Charlotte Remé administered a battery of visual and retinal sensitivity tests to a group of patients who had used lithium for up to twenty-two years. In almost all cases, sensitivity was reduced compared with normal levels, even though there was no associated eye disease. Although we know that lithium acts at various levels on the genetic machinery of the inner clock, it is clearly also serving to restrict the eye's response to light input. In that sense, lithium may act as "pharmacologic sunglasses"—mimicking the simpler environmental effect of staying in a darkened room or wearing blue-blockers.

If lithium is acting like pharmacologic sunglasses—or a reverse alarm clock, helping people realize that it is time to go to bed, as opposed to getting up—that might explain why it has been so successful at keeping patients with bipolar disorder stable. Increasingly, clinicians and researchers are realizing that sleep loss is both a precipitant of hypomania and a symptom of it. A new analysis of bipolar disorder even views the illness as a combination of four core dysfunctions, one of which is a disturbance of circadian rhythms interfering with sleep patterns.

Depression often involves changes in sleep pattern as well, whether insomnia or hypersomnolence (sleeping longer). Those with SAD often sleep two to five hours longer than normal. Even though most of us sleep more as the nights grow longer in fall, the increase is more extreme in SAD. Once the depression hits, the urge to sleep is reinforced. Some patients call it a way to "escape from the world." (Patients with nonseasonal depression often say the same thing.)

Even with all that sleep, those with SAD do not feel refreshed when they awaken. They feel sluggish all day and often go through major afternoon slumps. Of course, slumps in themselves are not a sign of depression. Many of us who are not depressed also feel afternoon slumps, and maybe an urge to nap, that intensify in winter. But patients with SAD are hit harder.

Another symptom of clinical depression is changes in appetite, in either direction. Those with SAD are likely to eat more compared with their summer habits. Again, this is somewhat true for the population as a whole, but patients with SAD are hit much harder. Fall brings a craving especially for carbohydrate-rich foods, and patients can easily gain ten to twenty pounds, or even more. Luckily, in the late spring most or all of this excess weight is easily lost as taste preferences change and physical activity resumes.

People who are feeling so down are likely to lose the urge to interact with others, too. They may do what they can to appear normal to those around them, especially at work, but inside they feel miserable. Keeping up this kind of front requires a major effort and usually ends up making them feel even worse. Meanwhile, at home, family members and friends are likely to notice the changes and become concerned. What happened to family meals and evenings together? Why do phone calls from friends go unanswered and unreturned? What is it with coming home from work and going straight to bed? As for sex, it becomes perfunctory and only at the partner's insistence, or vanishes for the duration. As one male patient put it, "I become sexless in winter—in summer I'm rarin' to go!"

The cumulative mental burden of low mood, sleep and appetite changes, pressing fatigue, reduced productivity, and social isolation can lead to a grinding sensation of being ill at ease, made worse because there does not seem to be any specific external cause. This "psychic anxiety" is pervasive in SAD . . . unless things get so bad that the motivation to cope is lost entirely, leaving only lethargy.

As many as one hundred million Americans, about a third of the population, suffer from noticeable seasonal swings. About twenty-five million are laid so low that they cannot keep up their normal activities. It is they who would receive a diagnosis of SAD, which by definition involves a moderate-to-severe clinical depression. The other seventy-five million show the same pattern of seasonal mood, energy, appetite, and sleep changes, but in a less severe form—winter doldrums rather than winter depression. They are better able to get through the day without drawing unwelcome notice from coworkers, but it costs them a constant effort to conceal their pervasive sadness and fatigue.

Calendars and Latitudes

Seasonal change is not the only factor that promotes SAD and winter doldrums. In mountain valleys, it can seem to take forever for the winter sun to rise. In regions known for their constant cloudiness, it can seem that the sun doesn't even bother to rise. In densely built cities, the apartment house across the street can keep you from knowing whether the sun rose or not. Anything that cuts down on your access to outdoor morning light raises your likelihood of becoming depressed during the winter season.

The most pervasive factor, the one that affects everyone for better or worse, is geographical latitude. As those in real estate say, location, location, location. In winter the sun rises later and later, the farther north we go. We can see this clearly if we look at New Year's Day, which comes during the time of year when the sun rises latest. In Brownsville, Texas, way south on the Mexican border, sunrise comes at 7:17 AM. That same day, it rises at 7:42 AM in Topeka, Kansas, in the center of the U.S. In Fargo, North Dakota, up near the border with Canada, it doesn't come until 8:12 AM. In Fairbanks, Alaska, the sun finally puts in an appearance at 10:54 AM on New Year's Day!

The statistics for SAD and the less severe winter doldrums follow a similar trend, but only up to a point. The number of cases is fairly low in the far southern areas of the U.S. and gradually rises as we move up into the middle states. At that point, however, it levels off. All else being equal, your chances of being affected by a seasonal disorder are roughly the same whether you live in Colorado, Kentucky, Montana, or Maine. This is an example of what scientists call a *ceiling effect.* In commonsense terms, most of those who are susceptible to developing SAD will develop it if they live somewhere with winter sunrises of 7:30 or later. How *much* later does not seem to matter. It is interesting to note that throughout Europe, which lies farther north than the U.S., sunrise on New Year's Day comes later than 7:30 AM whether you are in Rome, Italy (7:38 AM) or Stockholm, Sweden (8:44 AM). And from south to north in Europe, the likelihood of getting winter depression stays about the same. This provides further evidence for the existence of a ceiling effect.

One other note: At the equator, the sun rises at about 6 AM throughout the year. According to our research, there is no SAD there. No winter—no winter depression.

Moving and Mood

Could a SAD sufferer in Michigan get relief by spending the winter much farther south, say in Georgia? It's a possibility, even though SAD is found, if at a lower rate, in Georgia, too. Many SAD sufferers who can afford to get away know that they get quick relief on winter vacations in the Caribbean or Mexico. The problem is that when they have to return home a week or two later, they may crash to an even lower level than when they left, if it's still midwinter. If the vacation is timed for late winter, however—say in late February—the winter cycle can be stopped dead in its tracks, with an earlier "springtime remission" than if they had stayed home.

Moving between regions can have a negative impact, too. Consider those who immigrate to New York from the Caribbean. Before the move, they may have had no history of depression, only to experience recurring SAD starting their first winter in the North. In contrast, immigrants from Scandinavia who are very well acquainted with SAD may find permanent winter relief when they move to New York. It's all relative.

What is the best place for you to live, if you feel you are vulnerable to SAD? The most accurate (if not very satisfying) answer is, it depends. A major move south might help—or it might not. We had one patient who moved from New York to Atlanta and never again had a problem with winter depression. Another, who moved from New York to San Diego, slightly farther south than Atlanta, relapsed in winter, just as if she hadn't moved at all. As a result, she decided to move back to New York and control her winter swings with light therapy.

One individual factor that makes a difference is genetic heritage. Apparently, there are two related genetic tendencies that are linked to SAD. One affects susceptibility to depression, while the other affects sensitivity to changing seasons. Someone who has inherited both of these *and* who lives in an area with late sunrise during winter is a very likely candidate for SAD. Someone with only the seasonality tendency may sleep more and gain weight during the winter, but without becoming depressed. As for those with neither vulnerability—about half the population, according to our data—they generally get through the winter just fine.

Eleanor's Story—Part II

Eleanor knew that her seasonal depression was a real problem, and she wanted to find a permanent solution. "My grandfather had a store where he sold scientific equipment," she recalls. "Maybe that's why I have a bent to things that can be studied. I know how these things work, the lights, melatonin. It made sense in scientific terms, but it also made sense intuitively because of all those spring days when I felt better outside. Then at some point I called Columbia, and it led to Dr. Terman. I think it was fall. I learned from him that fall is a dangerous time, because there are quick changes in the light.

"He suggested I begin cautiously, seven thousand lux in the middle of the day," she explains. "He was concerned that I start carefully because I was already agitated and unable to sleep. That's very unusual—I usually oversleep. I was so agitated then, I couldn't lie still or sit still. That feeling was new. It was a different feeling. It was a combination: My mood was depressed, but physically I felt energetic and agitated. I just knew something was off."

An incident that helped convince Eleanor that the light box was working for her came when she forgot to order extra bulbs for it. "The bulbs were still working even when they were dimming. The light didn't look any different," she recalls. "I had to order new bulbs, and I had to wait for the shipment. But within a day or two, I was feeling better. That was very convincing, and very dramatic—look what happens when I don't use them, and when I do use them.

"I'm using the light box year-round now. In the beginning I had anticipated using them just from October or November until about April. But now I use them all year and I increase the amount I'm using as winter starts. I was looking for something that would address the biology without using medications. There's always a risk of side effects, but I feel light therapy is more benign and less risky. And you get the benefits much more quickly! And you feel more in control than you do with meds, because you can make adjustments quickly for yourself."

Asked if there was anything else she wanted to communicate to our readers, Eleanor said, "I've been thinking about how skeptical people feel. Part of it has to do with depression. You feel hopeless, pessimistic, it's hard to believe there's a way out, that there are things which are effective. With something like the light box, it's more challenging because you do have to make room for it in your life. You have to rearrange your schedule. It's not mainstream, and it's not as simple as meds. Combine that with the hopelessness that depression creates—it's a big bar to get over."

She adds, "I have a cousin who is struggling with depression. I suggested he give it a try. I bought one [a light box], and I thought if it didn't work for him I would use it as a demo for my practice. He was skeptical, but it's worked! It's also worked for his cat," she says with a laugh. "His cat is very grumpy. But when he has the lights on, the cat purrs and nuzzles and isn't grumpy at all."

SAD and the Medical Profession

It happens all the time. A patient comes in and tells the primary care doctor, "I've been feeling so tired lately. Do I have something?" What we often mean by this is that we are suffering from symptoms of depression, whether seasonal or not. We don't put it in those terms, though, for many reasons.

- We may think that a physical problem is more easily treated and cured.
- We may think of doctors as experts on physical rather than psychological problems.

- We, along with our doctor, may not have made the connection between our problem and the time of year.
- We may simply not know what it means to be depressed.

Unfortunately, most doctors are not yet familiar with the facts about SAD. As a result, they don't ask the right questions or prescribe the right treatment. They are likely to advise better sleep habits, more exercise, and a healthier diet—all good things in themselves, but not solutions to the problem. And if they do decide that the patient is suffering from depression, they probably write a prescription for an antidepressant.

It is very easy for a physician to conclude that a problem was trivial when the patient returns a few months later and the symptoms are "gone." The symptoms, yes, because spring has arrived and the depression has lifted. But SAD doesn't disappear in summer. As a syndrome it includes both winter downs *and* summer ups. An informed doctor who asks the right questions could diagnose it even when the patient is feeling fine in spring or summer. But what clinician—and what patient—is going to consider the overall annual pattern when things are going well in summer?

So it is up to the patient to take the initiative. Otherwise the necessary discussion may never happen. If you are concerned that you may fit the picture we have given for someone with a seasonal disorder, a constructive first step is to take the free online seasonality questionnaire (see *Resources for Follow-up*, p. 301). You'll get immediate, confidential feedback about how closely your experience fits the SAD profile and whether it appears serious enough to warrant a clinical consultation. If the answer is yes, print out the personalized report and tell your doctor you need an appointment to review the information together.

6. Geography and Time

Seasonal changes are not the only source of tension between internal and external time. Yes, the longer nights of winter lead to conflicts between our internal clock and external time. When the sun rises later, our circadian rhythms tend to drift later as well. At the same time, we are expected to go on sleeping on a standard schedule. This discrepancy between the clock in our brain and the clock on the wall readily leads to mood and sleep problems as well as morning grogginess. Other aspects of daily life can also have a serious impact. One, only recently recognized, is the particular place you live within your time zone. Another arises from the legally mandated custom in much of the world of changing the clock twice a year to and from daylight time.

Time Zone Disruptions

Until the middle of the nineteenth century, all time was local time. Noon arrived when the sun was at its highest daily point in the sky or when a church bell or factory whistle signaled it. But the spread of railroads made it essential to standardize time, and the telegraph made it possible, as we discussed in Chapter 1. Eventually, we arrived at our current arrangement of time zones. This carves up the globe into segments, shaped roughly like orange slices, which generally have an hour's time difference from the next ones over to the east and west. The geographic coordinate system that specifies east-west positions is longitude. During the course of a twenty-four-hour solar day, the sun appears to pass over all 360 degrees of longitude. In an ideal world, then, each time zone would be 15 degrees wide (360 degrees divided by 24 hours).

The world we actually live in is not quite ideal, however. The lines of longitude that geographers drew run ruler-straight, but the borders of time zones owe as much to politicians as to geographers. State and national boundaries, rivers, mountain ranges—almost anything can be an excuse to fiddle with time zones. In Indiana, the state legislature decided that a large swath of land that belongs geographically to the Central time zone should be in the Eastern time zone instead (to stay in sync with Wall Street!). And some parts of the world brush aside the notion of time zones altogether. China, for example, uses only one time zone for the entire country, even though its east-west expanse would entitle it to have four or five different zones. The result is that in China's far western areas, the sun may not arrive at its highest point in the sky until 3 PM! In response to this, the workday in these areas generally starts an hour or even two hours later than in most of the country.

A Report from the Front

Sue Dillon of Carmel, Indiana, is president of the Central Time Coalition. The goal of this organization is "to promote the allocation of available sunlight and darkness in a manner that provides the greatest peace, safety and well-being to the citizens of Indiana" by returning the state to "its geographically correct Central Time Zone."

Dillon became deeply involved in the issue after a terrible incident in her town. "In January 2009 a local fifteen-year-old student slipped and fell under the wheels of his school bus," she explains. "This accident occurred at six forty-five AM, and the boy wasn't found for another twenty minutes. That pushed me over from writing my annual letter to the editor to becoming an activist. This student and his family had emigrated from China two years prior to the accident. He was an only son. They came to Carmel because its school system is one of the best in Indiana and they wanted the best for their son. Now he is dead and morning darkness was a major contributing factor."

This is just one of many cases in which schoolchildren have been injured or placed in danger on their way to school in winter morning darkness. "The one that caused the most community trauma occurred last year in October," Dillon says. "A twenty-seven-year veteran teacher hit two boys walking along the road on their way to school. One boy was killed and the other had broken vertebrae. I'm very angry that perceived profits have motivated business interests to influence adoption of a time zone that places our 1.35 million students (21 percent of our population) at a safety risk every morning they have to travel to school in the dark. Eastern time is the wrong time for Indiana."

Even when time zones fit the model of being fifteen degrees wide, sunrise at the western boundary comes a full hour later than at the eastern edge. This may not sound like a lot, but an hour's difference in sunrise can have marked biological consequences. If, like so many people, you are locked into a nine-to-five workday, you may have to wake up at 7 AM to get to work on time. If you live on the eastern edge of your time zone, the sun may already be above the horizon when you get out of bed, depending on the time of year. If you live on the western edge, however, you may have to get up while it is still dark outside.

Erin's Story

Erin D. is forty-six and married, with four children ages thirteen to twenty. Her husband is a physician, and she has her own business as a career counselor. "I can do this from home," she says, "which is great, because my primary job for the past twenty years has been raising my kids." The family lives in a small town outside Indianapolis, Indiana. "I actually moved to Indianapolis when I was five, and have lived in the greater Indianapolis area ever since."

Like most of Indiana, the area where Erin lives uses Eastern time, even though it is geographically closer to the Central time zone. This is a very sore point with her. "The majority of Indiana is like a chunk cut out of the 'true' Central time zone that has been placed in the Eastern time zone," she points out. "I have sworn for years that I would move to Florida or California if given the chance. I get so tired of Indiana winters and just feel like I have to escape. I wake up in the dark and it is dark for almost two hours after I get up. Then it is dark before dinner. Add the ice and snow of winter and it's like we're trapped."

Her biggest concern during the winter is whether her children are safe on their way to school. "It is still dark until almost eight here. My youngest son is dead asleep still at that time, yet he must get up by six-fifteen to catch the bus at seven. He is grouchy and complains every morning. It is so hard to get him moving. It seriously looks like midnight when he heads out to the bus. He has to walk all the way to the end of our street and I worry every time that the cars won't see him. The middle school starts at and that means that students are waiting for the bus in the dark most days of the year. It is awful. When it snows, there is no place for him to walk except the middle of the street if the streets have been plowed. Cars and buses can't stop fast enough at the last minute, if they even see kids in the dark."

The situation is no better for her daughters, who are in high school. "The eldest, who is a senior, must drive herself and her sister, a sophomore, to school. A couple of years ago they took the bus to school, but now they drive because the new bus stop is over a mile away. It is too dark and too dangerous to try to walk a mile to the bus stop with backpacks and no sidewalks. More students are driving to school now than ever before. These are all inexperienced drivers who are out in full-on traffic situations in the dark (and often the dark and ice), and I would venture to say they're usually not fully awake."

Summertime brings its own problems, as Erin explains. "These past few summers we have not felt ready to eat until about eight PM. Somehow, it's hard to feel hungry for dinner when the sun is still high in the sky. It throws our whole day off. We eat late, go to bed late, get up late. Combine the wrong time zone with daylight saving and we are effectively on double daylight saving time. Now we are two hours out of sync in the summer, instead of just one. The 'sunburn time' of ten to three is really twelve to five or so. We don't have 'true noon' at noon, but rather at one forty-five. The evenings are *hot*, and by the time they cool off, it is very late."

Erin's sister and her family, who live in Manhattan, notice the difference when they come for summer visits. "My sister and her husband always say how crazy it is that it is still light here at nine forty-five PM. Their summers [in Maine] are punctuated with marshmallows over fires and lightning bugs and all the things that kids remember about their summer evenings. My kids haven't seen lightning bugs for five or six years because they don't come out until ten PM. I don't go to fireworks celebrations over the Fourth of July because they don't start until ten PM.

"I hate knowing that things have been artificially adjusted," Erin adds. "I can see absolutely *no* reason, not even one, for us to be on Eastern time. We're out of alignment geographically. Our days could be starting with bright walks to the mailbox and brighter light for the school bus and drives to school, but they aren't. Most mornings I think about how virtually every other city has the sun shining while we're still in the dark.

"Being in the Eastern zone makes no sense," Erin concludes. "We're lighter longer in the summer evenings by about an hour (on already long summer evenings), and we're darker an hour longer in the winter mornings by about an hour (on already dark winter mornings). How can that be healthy?"

Consider two cities, Boston and Detroit. Both are in the Eastern time zone, though Detroit is about seven hundred miles west of Boston. If we look at April Fools' Day, we find that the sun rises in Boston at around 6:30 AM, but in Detroit it isn't up until 7:15 AM. This is a very noticeable difference if you're getting up at seven.

We recently carried out a Web survey of thousands of people who answered questions about their seasonal mood changes. The responses were anonymous, but we were able to use postal zip codes to pinpoint the home latitude and longitude of those who responded. We separated out people in the northern half of the U.S., where the most cases of seasonal affective disorder occur, and sorted their answers by different time zones. Within each time zone, the farther west people lived, the more likely it was that they showed signs of SAD. Then we compared responses from either side of the boundaries *between* time zones. There was a sharp drop in SAD between those at the western edge of one zone and those, often just a few miles away, on the eastern edge of the next time zone over.

This longitude effect is amplified by the way winter nights get longer as we move to more northern latitudes. Winter sunrises in the lower forty-eight states of the U.S. can come as late as 8:43 AM. This is a bleak start to the day for the average worker. Consider places such as the western extension of Michigan, which is in the Eastern zone; most of North Dakota, in the Central zone; and about one-third of Montana, one half of Idaho, and a peculiar patch of eastern Oregon, all in the Mountain zone. According to our statistics, these are the areas where winter depression hits hardest, because of the double whammy from the natural sunrise-related latitude effect and the time-zone-legislated longitude effect. (Alaska, far north *and* far west, is in a category by itself. In Nome, the sun in winter sometimes doesn't come up until after noon!)

On the brighter side, you may be lucky enough to live in an area at the eastern edge of your time zone. These include Alabama, Tennessee, and Kentucky, which lie along the eastern edge of the Central time zone; eastern New Mexico in the Mountain time zone; and the eastern stretches of Nevada and southeast California in the Pacific time zone.

It is odd to realize that you could essentially reverse the effect that the delayed winter sunrise has on your biological rhythm by crossing a time zone border. And given the irregular ways these borders are drawn, you might not even need to move east or west. In the dead of winter, the sun rises just after 7 AM in Wells, Nevada, but about one hundred miles due north, in Twin Falls, Idaho, it isn't up until after 8 AM. (It really rises at the same time in both locations, but only if you are going by solar time.)

Of course clocks and schedules are essential to our structured world economy and culture. That means we are stuck with time zones, however difficult they make it for some of us. Even if the world were divided into evenly spaced time zones, each exactly fifteen degrees of longitude wide, the longitude problem would not disappear. Those on the western edge of their zone would still be more susceptible to SAD, because the later winter sunrise deprives their inner clock of the early-morning light signal it expects and needs. In principle, they could deal with this by moving to the eastern edge of the next time zone—and if someone is battling a devastating depression each winter, they might even do so. Some who live in the western part of their zone and are lucky enough to have flexible schedules can simply try waking up later. For most of us, though, a more practical course is to give careful thought to the therapeutic options that are a major reason we wrote this book.

Daylight Time

It happens twice every year: the diagrams of clock faces with arrows and phantom hands, the newscasters' reminders to "spring ahead, fall back," the scramble to reset all the various clocks and timers around the house, and all too often, the alarm clock you forgot to change going off an hour early or an hour late. Most of us accept the switches in and out of daylight saving time with only an occasional grumble. But it can take up to a week for our bodies to adjust, and for some people the transition is quite jarring. It probably doesn't even occur to us to think that, like time zones themselves, daylight saving time is an arbitrary, legislated tinkering with time.

This ritual of changing the clocks near the start of spring and again in the fall dates back to the early years of the twentieth century. At the time, a primary use of electricity was electric lighting in homes, offices, and streets. Those who supported the new system argued that by taking a daylight hour away from the early summer morning, when most people were still asleep, and tacking it onto the evening, when most people were still up, the need for artificial light would go down sharply. When coal became critically scarce during World War I, the warring powers in Europe accepted this argument and adopted what Europeans still call Summer Time. North America soon joined them but named it daylight saving time instead.

Whatever the shift is called, it is a semiannual source of widespread confusion. Just what is it that is springing ahead or falling back—the clock, me, or both? Are we gaining an hour or losing it? The day after we have to change the clocks, will the sun come up earlier or later? Predictably, the shift confuses the circadian clock in the brain as well. This inner clock, as we know, relies on exposure to light, especially natural light, to keep itself in sync with the daily cycle. But the sun does not understand about shifting to or from daylight saving time, and neither does your inner clock.

Let's say you live somewhere near the middle of the continental United States—Kansas City, Missouri, for example, in the Central time zone. And let's say that you have a daily rhythm that gets you to sleep at 11 PM and wakes you up at 7 AM. In 2013, daylight saving time begins on Sunday, March 10. On Saturday, March 9, you wake up as usual at 7 AM (CST). The sun is already up and has been since 6:39 AM. The next day, after the time change, your inner clock awakens you at what for it is the usual time. Once again, the sun has been up for over twenty minutes. Then you look at the clock on the wall. Assuming you remembered to change it the night before, you are startled to see that it is already 8 AM Central daylight time, not 7 AM. Come Monday morning, you still have to be up by 7 to get to work on time. But now you will have to wake up an hour earlier than your inner clock is prepared for, and you will wake up to darkness, more than half an hour before the sun comes up. And if you also lived in the western part of your time zone, it might *still* be dark when you leave for work and the kids have to go wait for the school bus.

How does this fiddling with time twice a year affect people's health and well-being? For some, quite badly. A recent study published in the *New England Journal of Medicine* found that during the week after the spring shift to daylight saving time, the rate of hospital admissions for heart attack went up by as much as 10 percent. The researchers suggested that this was because time change in spring costs people an hour's sleep. They still have to wake up to get to work on time, but the time they have to wake up is objectively an hour earlier. It is interesting to note that retired people didn't show this increased risk of heart attack. This may be because they aren't forced to wake up an hour earlier when the clocks change. According to this same study, the shift back to standard time in the fall had less effect on heart attacks, which the researchers credited to the extra hour of allowable sleep. Of course, that extra hour also allows the internal clock to drift later in accord with its underlying cycle longer than twenty-four hours. So, relative to the springtime change, the adjustment is physiologically easier.

Young people are also affected badly by the switches in and out of daylight saving time. A recent study compared high school students in parts of Indiana that use daylight saving time with other parts of the state that stay on standard time year-round. On average, those who were subject to daylight saving time had scores on the Scholastic Aptitude Test (SAT) more than sixteen points lower than those in areas that didn't change the clocks. That is a startlingly large difference. Given that there is a well-documented link between SAT scores and earnings in adult life, the researchers calculated that daylight saving time is to blame for a loss of billions of dollars a year in potential earnings just in Indiana!

For most of us, adjusting to the time changes doesn't seem to cause more than a week or so of disruption. How much disruption depends partly on whether we are "springing ahead" or "falling back." The circadian clock finds it easier to adapt to falling back into standard time. After all, most people's normal clock cycle is longer than twenty-four hours, perhaps twenty-four hours and twenty minutes. If they leave their blinds down and catch an extra hour's sleep on the Sunday morning of the time change, their inner clocks will naturally drift toward being in sync with the change. The spring change is harder. It requires the inner clock to make a correction of a full hour, in addition to the daily correction of twenty minutes or so that it usually needs to make. Worse, it has to make it when the necessary morning light is coming an hour later. Some researchers even conclude that our clocks *never* totally adjust to daylight saving time. A survey of fifty-five thousand people in Central Europe indicated that, across the year, sleep patterns on free days were closely linked to the time of dawn under standard time but lost this linkage during the months that daylight saving time was in effect.

As of 2007, daylight saving time in the United States was extended by four weeks. It now begins on the second Sunday in March and ends on the first Sunday in November, a stretch that takes in almost two-thirds of the entire year. This is hard on a lot of people, and particularly hard on those who are battling winter depression. The loss of an hour in March means that according to the clock, sunrise now comes as late as it did at the end of December, during the darkest days of the year. Nor are we likely to gain back the lost morning light in November, when we change back to standard time. Yes, the sun comes up an hour earlier by the clock, but even so, many of us will still be waking into near darkness.

Some numbers may help to make these points clearer. We earlier used Kansas City in 2013 as an example. After the shift on March 10 to daylight saving time, the sun rises at 7:38 AM. That is the same time (by the clock) as sunrise back in late December, its latest time in the entire year. As for the change on November 3 back to standard time, that Saturday the sun comes up at 7:47 (Central daylight time), and on Sunday, after the shift, at 6:48 (Central standard time). If you have to get up at 7 AM on Monday, you will find the sun is up before you, but not by much. And two weeks later, you will once again be getting up before the sun.

If your experiences lead you to think that the shifts into and out of daylight saving time have a negative impact on you, there are some steps you can take to reduce that impact. In the fall, the easiest way to deal with the transition is simply to keep the blinds drawn on the Sunday of the change and try for an extra hour's sleep. This will encourage your inner clock to drift later, putting it on course to become synchronized with standard time.

The change in March into daylight saving time is more jarring because our inner clock needs to shift *earlier* to accommodate to the change. Don't wait for the day of the time change and begin a week or more of suffering. Instead, plan a week ahead. Starting on the previous Sunday, set your alarm ten minutes earlier on each successive day *(no snooze button, please!)*. Get up, turn on the room lights, and open the blinds. If you know from experience that the time change is very hard on you, you can further ease the transition by sitting at a bright light box for ten minutes after you wake up. (Chapter 7 explains how to use a light box.) During this adjustment week, get to bed earlier as well, even if it means skipping the eleven o'clock news. By the Saturday of the time shift, you'll be waking up an hour earlier than your customary time. This may feel strange, but it means that your inner clock will already be in sync with daylight saving time when it begins the next day.

PART 3

Interventions

7. Healing Light

If you are getting too little light, or getting enough but not at the right point in the day, you are at risk for a long list of problems. Fortunately, the cause implies a cure. If getting too little light is the cause, getting enough light and at the right time offers a way of dealing with all these problems. By restructuring the amount and pattern of our daily light exposure, we can achieve remarkable therapeutic effects, even in cases of severe depression or sleep disorders. Best of all, the intervention is environmental, not pharmaceutical. Results often come more quickly than with drugs, sometimes in days. And if it turns out that light therapy is not enough, drugs can always be added or substituted.

For most people, the hardest part of starting bright light therapy is getting used to a new morning habit. But that does not have to be any more of a hassle than brushing your teeth, taking a shower, and getting dressed for the day. You will not be obliged to sit there staring at a bunch of lights. You can spend the time productively. You can read the paper, phone friends, check your e-mail, plan your day, or simply enjoy a more leisurely breakfast. If even this turns out to be too much to manage, there is also a new alternative available. Dawn simulation therapy automatically brings a sunrise to your bedroom each morning as a selected wake-up time draws near. We describe both methods in this chapter.

You can use bright light therapy at any point in the day to increase your energy and alertness, and many do. As one patient said, "It's better than coffee!" Its more profound use, however, is as a delicate tool for adjusting your circadian rhythms. When you expose yourself to therapeutic levels of light toward the end of your internal night, the neural signal that is transmitted to your inner clock causes it to shift earlier. This is similar to what happens normally as winter changes to spring and the sun rises earlier. The inner clock resets, the pattern of melatonin production changes, and the problems of sleep timing, seasonal depression, or both, lift.

When we use the expression *bright light therapy*, what are we talking about? How bright? Most of us probably think of the brightness of artificial light in terms of the number of watts of electricity a lightbulb uses. The bulb in the refrigerator is 25 or 40 watts, and the one we like to read by is 100 watts or even brighter. This has become a lot more confusing recently, as more of us switch from traditional incandescent bulbs to high-efficiency compact fluorescents and LED lamps, which use fewer watts to give the same level of light. In any case, what matters to our physiological system is not how much electricity a bulb uses or how bright it looks, but how much light actually reaches the eyes. This is determined by a combination of the output of the light source and the distance from the source to your eyes, which can be measured in lux.

During the day, the amount of light that reaches your eyes inside an average home or office is usually between 50 and 300 lux. This is about the same as if you were outside at twilight. But in the period soon after awakening, our nervous system *expects* full daylight, which is more like *10,000* lux—hundreds of times as much. How bright is 10,000 lux?

Imagine that it's a clear morning, about forty minutes after sunrise. You are taking a walk along the beach, not directly facing the sun. Oh—and no sunglasses. Your eyes are probably receiving around 10,000 lux. Needless to say, not many of us get that much light that early in the day, if at all. In today's urban society, far too many of us spend practically all our waking hours in what amounts to perpetual dusk.

We can make up for this deprivation by using a therapeutic light box. You may be thinking, "Oh, sure, a light box. My doctor uses one to read my chest X-ray. And when I was a kid, there was one in art class that we used for tracing designs. What's the big deal? Isn't it just a box with a light inside?" Partly, you are right. The basic structure of a therapeutic light box *is* simple and obvious. It's a box, a light source inside, and a diffuser on the front to keep you from being forced to stare directly at the light source. But the light boxes doctors and artists use are designed to illuminate translucent objects, such as film or tracing paper, from behind. In contrast, therapeutic light boxes need to provide a particular high level of light to the eyes, without causing harm or distress. Their design involves some subtle features that have called for a great deal of thought and experimentation to get right.

Light Intensity

The first time you turn on a light therapy box will probably be the first time you experience light of full outdoor daytime intensity inside your house. You may feel that you are already getting plenty of light because you like to bask by a bright window while the sun is high in the sky. Basking is fine, but you are unlikely to be getting anything close to 10,000 lux. As for the early morning, when light therapy has its greatest effect, the sun is not yet going to be high in the sky. For much of the year, it may not be up yet at all.

Ah, you say, but I don't need a light box. I live in a really sunny area. Good for you, but that will not help you much if, like most of us, you spend most of your time inside. A fascinating study was carried out in San Diego, one of the sunniest cities in the United States. For a week or more, a group of young, vigorous medical students wore battery-powered devices that continuously recorded light levels. The result? Hardly any of them got more than twenty minutes of outdoor-level light during the course of an entire day, and certainly not in the period just after getting up. There was plenty of light available to them—light levels on San Diego's Mission Beach can easily top 100,000 lux at noontime—but they spent the time indoors in 300 lux or less. This is not just a depressing thought, it is a depressogenic situation, one that *causes* depression in vulnerable people.

All light boxes are not equally effective, either. Far from it. The earliest therapeutic light boxes, back in the 1980s, provided about 2,500 lux, the level outdoors about a minute after sunrise. Patients had to sit in front of them for two hours or more a day to get the antidepressant effect. When we upped the level to 10,000 lux, the daily sessions dropped to half an hour. This made light therapy a realistic possibility for people who need to leave the house for work. Unfortunately, it is not always easy to find out how much light a commercial device really provides. This poses big problems for consumers and patients who are forced to rely on the specifications listed in an ad or an online description.

Sometimes these specifications are less than accurate. Manufacturers who want to claim that their device provides 10,000 lux can simply place the light meter behind a narrow tube and aim it at the center of the light box. This will guarantee a high reading, far higher than if they measured the open field of illumination, which is what your eyes pick up. Under proper laboratory conditions, lux levels are measured with a broad, rounded probe placed at the same distance from the light box that you sit at. Anything less precise and you do not know what you're getting.

This is not the worst of it. Some suppliers manage to sell light boxes at very low prices by simply fudging the numbers. After all, how many buyers are going to get ahold of an accurate meter and try to verify their claims? It is as if you filled a prescription for 40-milligram Prozac and discovered that while the label said 40 milligrams, the tablets inside were only half that strength. If that happened with a medication, those responsible would face serious criminal charges, but to date there is no regulatory control over light box specifications. We have raised this issue with agencies including the U.S. Food and Drug Administration and the Federal Trade Commission, but so far there has been no response. Short of regulatory standards, the nonprofit Center for Environmental Therapeutics advises consumers and doctors about the features they should look for in a therapeutic light box, as described in the following sections.

The Angle of Light

Think about taking a walk outside. Where does the light come from? Except when the sun is very near the horizon, toward sunrise or sunset, we receive the light signal from 30 degrees or more above our line of sight. We are not able to tolerate a very bright light that comes straight toward our eyes. That's why cars come equipped with sun visors, and probably why we come equipped with eyebrows. Yet some of the light boxes on the market pay no attention to this obvious fact. They are designed in a way that places their light source directly at eye level.

Just as hard on us is light that arrives from *below* the line of sight. This has the same effect as walking across a snowy field on a sunny day. Unless you are wearing a serious pair of dark glasses, the reflected glare from the snow is guaranteed to have you squinting or turning away. You'll be lucky if you escape getting a headache from the tension in the muscles around the eyes. And again, some commercial light boxes ignore this factor. If you search online, you will easily find examples of "therapy lamps" built to rest on a table or counter, below eye level.

This neglect of basic design factors is more than unfortunate. It means that people who have decided they may benefit from light therapy and have taken the step of buying the equipment they think they need are likely to experience disappointment. If the position of the bright light source makes you squint, or look to the side, or close your eyes to reduce the glare, you are probably missing the therapeutic benefit. Worse, you are more likely to quit light therapy altogether and warn off anyone you know who might consider it.

The solution to this problem is simple, if too often overlooked. Our clinical trials of 10,000-lux light therapy make it clear that an adequate light box should mimic the experience of sitting outdoors without staring up at the sky. This means placing the light source on a table stand in such a way that the light comes from distinctly above your line of sight. When it does, most people find that even a very bright light is comfortable, energizing, and even soothing.

What Color Light?

If you hang a crystal prism in a sunny window, you will see a rainbow of colors projected onto the wall or ceiling. White light contains all the colors of the visual spectrum. Studying that rainbow, we can see, from the longest to the shortest visible wavelengths, red, orange, yellow, green, blue, indigo, and violet. (Students from elementary school on use the name of ROY G. BIV to help them remember this order.) Each color corresponds to a narrow band of wavelengths in the spectrum. Complex colors such as brown and purple are formed by combinations of these elements.

If you have a recent camera, it probably has a built-in feature called "automatic white balance." That is needed because white light itself comes in different varieties, based on the balance of wavelengths in the mix. This balance is expressed as *color temperature*, measured in kelvins (K). When you buy lightbulbs, you are asked to choose the particular sort of white light you are after. "Soft white," around 3,000 K, is called soft because it is very easy on the eyes, producing minimal glare. The reason is that it has fairly low levels of short-wavelength violet and blue. A bulb that produces a little more blue will be labeled "cool white" and measure about 4,000 K. Then we have so-called "daylight" bulbs, at about 5,000 K.

As the amount of blue in the balance continues to increase, we get to "full-spectrum" light (a misnomer), around 6,000 K. Back in the 1980s, the earliest light therapy boxes were designed around full-spectrum bulbs. Researchers assumed that something more like skylight would be more effective. At the time, these bulbs were quite expensive, which added a certain mystique to the procedure. An echo of this can still be found today in the claims made for some light boxes.

The simple fact is that there is no particular advantage to full-spectrum light, as compared with lower color temperature white light. It contains more blue than the white light, but there is no evidence that it provides greater therapeutic benefit. It also produces more ultraviolet radiation than lower color temperatures, and this is something we definitely want to avoid.

Lately, we've seen some manufacturers push to new extremes, touting devices that emit an intensely bright bluish hue of up to 17,000 K. These are similar to the high-intensity headlights found on some expensive cars. This bluish light is really no longer a member of the "white" family. It is uncomfortable at best to sit in front of a bank of these lamps. Fortunately, there is no need to, since there is no evidence of any additional clinical benefit.

Careful clinical trials in our lab and elsewhere have clearly shown that any white light in the range from "soft white" to "daylight" (3,000–5,000 K) is equally effective for light therapy. The only discernible differences are that people have varied subjective visual sensations, depending on the color temperature. We do need a bit of blue in our white light, because blue is more effective in stimulating a response from the circadian system. But there is more than enough blue in warm or cool white light to do the full job. Anything more is simply gilding the lily and making the treatment intolerable for many users. The "full spectrum" ballyhoo is flagrant enough. The much-hyped boosts to 8,000 K and even 17,000 K are unforgivable.

During a light treatment session, we want visual comfort. We want to be able to concentrate on our breakfast or laptop or newspaper without glare aversion and headache. And if we're using a light box during the day, at work, the last thing we need is annoyance and distraction while we're trying to concentrate and be productive.

Ultraviolet Radiation

All fluorescent bulbs—the most common type currently recommended for light therapy boxes—produce some ultraviolet radiation (UVR). The higher the color temperature of the bulb, the more UVR it produces. Consistent daily use of a light box that is not properly shielded risks more cumulative exposure to UVR than people normally encounter in everyday life. Even small amounts of unshielded UVR can have a gradual effect leading to skin cancer, even though it may be years later. UVR can also harm the cornea of the eyes and predispose the aging eye to cataracts, though again the effect may not be seen for many years. In adults, UVR does not pass through the lens to the retina, but it does in children, creating a hazard. Inappropriately shielded full-spectrum light can also cause skin puffiness and burning within a few minutes. Importantly, the combination of UVR with many common medications can cause damaging photosensitization and what appears on the body's surface to be an exaggerated sunburn or skin rash. This combination can harm the eyes as well.

Because of these hazards, optimum UVR filtering should be an essential feature of any light therapy box, but the amount of protection that is actually provided in commercial units varies quite a lot. Almost any filter placed in front of a bank of bulbs will screen out some UVR. Even a pane of clear glass will cut down on UVR to some extent. The best available protection, polycarbonate panels, screens out more than 99 percent of UVR. But no matter what some advertisers say, no fluorescent light box can provide *absolute* "UVR-free" light.

One curious fact is that the earliest light boxes, in the 1980s, were designed around full-spectrum light sources that put out a lot of UVR. Some researchers believed that it was the UVR hitting the skin that was responsible for the antidepressant effect. One manufacturer even marketed a light box that featured a special UVR emitter along with the lightbulbs.

This idea that UVR was necessary to counteract depression was disproven by a number of controlled clinical trials. In one clever study, a group of volunteers sat at a light box with their bodies and heads fully covered to block skin absorption of UVR. Two slits let them see the lights. A second group had their upper bodies fully exposed to the light and UVR, but wore blindfolds. The verdict, as the investigators joked, was, "The eyes had it." Getting light but not UVR had an effect on depression; getting UVR with light only to the skin did not. In another study, one group wore UVR-filtering eyeglasses, while the other group wore apparently identical glasses that did not filter out UVR. There was no discernible difference in the effectiveness of the treatment.

But if UVR wasn't helping to counteract depression, why were some

SAD patients saying that they felt distinctly better from a session at the tanning parlor, which provides high doses of UVR? Some even became addicted to tanning, and still are. It is true that UVR can trigger the body to produce *endorphins*, the naturally occurring substances that have effects similar to opiates. And endorphins can generate a feeling of calm and elation. This is one of the reasons so many people find it relaxing to spend a day at the beach, under intense sunlight. The endorphin high is deceptive, however. The substance briefly masks the depression, but it does not address the underlying causes. Meanwhile the exposure to high levels of UVR is a hazard to health. Recently we are beginning to see legislative initiatives that bar minors from using tanning salons. A small step, maybe, but one that will benefit the next generation with lower rates of deadly melanoma.

Light Diffusion

We all know better than to stare up at the sun. It does not take much of a leap to figure that staring at very bright naked lightbulbs is not a good idea either. We need to put something between our eyes and the light source that diffuses the light and makes it less glaring. This is true both for fluorescent bulbs and the more recently developed miniature LED panels—an obvious point, perhaps, but one that has apparently eluded some light box manufacturers. Light boxes can make use of many kinds of diffusers, but some use none at all. If you try to use one of these, you find yourself looking at the light source directly or through a clear plastic shield. Not only do they make people squint and look away from the light, they also pose a hazard to the health of your eyes.

Light-diffusing panels come in two varieties, smooth and prismatic. You are probably familiar with prismatic diffusers, because they are often installed on the fluorescent ceiling lights found in offices. Up close, they resemble a field of tiny pyramids. On ceiling fixtures, they do an adequate job of diffusing the light but these are far from our eyes and we do not ordinarily look up at the fixtures. A prismatic diffuser on a light box is a different matter. The tiny pyramids project images of the lightbulbs behind them, which is hard on the eyes. In contrast, with a smooth, translucent diffuser the screen appears evenly illuminated, without glare. Why, then, do some light-box manufacturers use prismatic diffusers? We can think of two possible reasons. One, they cost less. And two, they transmit more light than smooth panels, so it is easier to reach an impressive level of light intensity.

Light Box Size

How big—or how small—should a therapeutic light box be? From the standpoint of convenience, we might say the smaller the better. A smaller device would be lighter, less intrusive, better able to fit in a backpack or carry-on bag. The problem is effectiveness. The goal of light therapy is to flood both eyes with the signal, especially in the periphery of the retina where you are not directly looking. That requires a field of illumination that is significantly bigger than small light boxes can provide. Miniatures may be more portable, but what good is that if they do not do what they are intended to do?

It is worth recalling that the earliest light boxes, which provided only 2,500 lux, measured two feet by four feet and weighed some forty pounds! In fact, they were standard ceiling units with extra bulbs added, set up on a table a few feet from the user. In contrast, today an effective, clinically proven light box weighs less than ten pounds, has an illuminated surface as small as twelve inches by fourteen inches—smaller than two side-by-side sheets of typing paper—and provides 10,000 lux. It also costs about $150, which is less than half of what the earlier devices cost.

Insurance Reimbursement

A therapeutic light box is not a prescription item, because it is not regulated by the U.S. Food and Drug Administration. However, prestigious professional groups, including the American Psychiatric Association, have endorsed light therapy after carrying out scholarly reviews of published clinical trials. As a result, most insurance companies in the U.S. will reimburse—fully or partially—the cost of a light box, provided the purchase is endorsed by a doctor's letter with an appropriate diagnostic code. Full reimbursement is mandated in Switzerland (and only for SAD), but not anywhere else. However, wherever you live, the chances of reimbursement are far higher than even just a few years ago.

Our clinic at Columbia has compiled a list of light-treatable disorders eligible for light box reimbursement, based on published clinical trials data or literature reviews. Insurance companies can accept or reject our judgment on this, of course, but in most cases they have responded positively. Your doctor may find it useful to review the list and provide an endorsement letter you can submit with your claim (see *Resources for Follow-up*, p. 301). These include subtypes of depressive, bipolar, and sleep disorders. By the way, you won't find SAD on this list under that name. The American Psychiatric Association's diagnostic manual calls it Major Depressive Disorder, Recurrent, or Bipolar Disorder, with Seasonal Pattern.

Constructing Your Own Light Box—*Don't!*

The physical elements of a light box are simple, so simple that if you are reasonably handy with carpentry and electrical wiring, you may be tempted to build your own. There are even websites that claim to provide do-it-yourself instructions. This is a very bad, and even dangerous, idea. The parts may be simple, but configuring them properly calls for professional skills. The light intensity, field of illumination, degree of diffusion, and wavelength composition (including UVR) call for calibration with specialized instruments. There is also the vital matter of electrical safety. We have heard disturbing reports of people who have burned their eyelids while sitting at their homemade boxes and of boxes that have gone up in smoke.

Bright light therapy is meant to improve your health, not put it at additional risk. That is the reason we urge doctors, patients, and consumers to be certain any light therapy box they select meets the criteria put forward by the nonprofit Center for Environmental Therapeutics. These are:

- Any light box you buy should have been tested successfully in peer-reviewed clinical trials or be demonstrated to be equivalent to peer-reviewed models. Manufacturers' claims are insufficient.
- The box should provide 10,000 lux of illumination at a comfortable sitting distance. Product specifications are often missing or unverified, which is a major downside of unregulated commercial promotion. The problem is resolved if you choose a device that has been used successfully in published clinical trials.
- Fluorescent lamps should have a smooth diffusing screen that filters out ultraviolet (UV) rays. UV rays are harmful to the eyes and skin.
- The lamps should give off white light rather than colored light. "Full spectrum" lamps and blue (or bluish) lamps provide no known therapeutic advantage.
- The light should be projected downward toward the eyes at an angle to minimize aversive visual glare.

- Smaller is not better: When using a compact device, even small head movements will take the eyes out of the therapeutic range of the light. To provide an adequate field of illumination, the rectangular screen size area should be at least 200 square inches.

To Self-Treat or Not to Self-Treat

Light therapy differs from drug therapy in that the device is readily commercially available without prescription. The treatment sounds simple, naturalistic—an obvious thing to try. And in many cases self-treatment succeeds, though we have some caveats:

- If you are already using one or more medications for depression or insomnia, don't add light therapy on your own. Your doctor should advise and monitor, because such combination treatment may require adjustments of drug dose.
- If you consider yourself only mildly affected by blue mood, lethargy, or difficulty getting to sleep, you may get quick results trying light therapy on your own. But don't rely on the treatment advice offered by manufacturers—they write their blurbs with the main objective to sell product, and they are not clinically qualified. First, take the online depression severity questionnaire to verify for yourself that your problem is run-of-the-mill and not clinically relevant (see *Resources for Follow-up*, p. 301). Then take the chronotype questionnaire to find out how to time your light therapy sessions at the start.
- If the depression questionnaire indicates that you should consult a professional, don't try light therapy on your own.

Light therapy may or may not be recommended, and you should let an expert help you decide how to address a serious mood disturbance.

- If your insomnia is so severe that you cannot fall asleep until after about 2 AM, don't self-treat with light. The chronotype questionnaire is not designed for people with delayed sleep phase disorder (as we described in Chapter 3).
- Be aware of potential side effects. When some people first start using a light box, they report headache, eye irritation, irritability, mild nausea, or early waking. This does not happen often, and when it does, it usually subsides within a few days. Even self-treaters may need to seek the advice of a clinician if side effects become severe. Skipping several days of treatment should tell whether the problem is light related or due to something else. The solution may entail reducing the amount of light exposure to test whether the disturbance subsides while the therapeutic effect comes in. Only rarely have users quit light therapy because of intractable side effects. Still more rarely, blue light—like ultraviolet light—can cause a photosensitizing reaction in combination with various prescription drugs and damage the skin or the retina. If you suspect such a reaction, see a dermatologist for starters. (For a list of drugs that may cause such a reaction, see *Resources for Follow-up*, p. 301.)

Dawn Simulation

When you switch on a therapeutic light box, the amount of light your eyes receive goes immediately from room level, say 300 lux, to therapeutic level, 10,000 lux. That is quite a jump, and it is very different from the gradual change we would experience outdoors at dawn. Does it matter? We asked ourselves that question as far back as the 1970s in our lab work with animals. Once we developed a system to mimic dawn and dusk twilights at low intensity, we found we could program the transitions for any time of day and night and guide the animals' circadian rhythms in either direction, later or earlier. The results clearly showed an influence on the timing of activity/rest cycles that differed from the pattern under standard light/dark cycles. The animals adjusted as they would outdoors—or when emerging from burrows in the field—with less of a shock reaction and more behavioral anticipation of the upcoming day or night phase.

Would something similar work for people with circadian rhythm problems or SAD? To find out, we built a light box with a set of mechanical blinds covering the front. The blinds were controlled by an early-model home computer that signaled a motor to open or close them smoothly in forty thousand steps over the course of one to two hours. Research volunteers agreed to have this ungainly contraption set up next to their beds. It might be mid-February outside, but inside the bedroom the light level gradually rose just the way it would if you were camping out in a tent under the open sky in springtime. After just a few nights, the volunteers were sleeping through the first light of dawn and waking up spontaneously when the level reached that of sunrise, around 300 lux. And their depression sometimes lifted in as little as a week.

One of our volunteers was Rosie F. She suffered from seasonal bipolar disorder, going from deep depression in the winter to super-high energy and euphoria in the summer. Her winter sleep pattern was all over the place, changing from day to day. She took long naps during the day and woke up for long periods at night. When she tried our primitive dawn simulator, she responded remarkably. She woke up regularly at the simulated sunrise, though she still went to sleep at variable times and took scattered naps. And while her depression scores improved, some residual symptoms remained.

We decided to try adding a dusk simulation. As she got into bed, the light gradually faded from 300 lux down to a very faint glow like starlight, around 0.001 lux. Within days, she was falling asleep about twenty minutes after the dusk fade began and sleeping through the night until simulated sunrise eight hours later. She stopped all napping and showed a complete remission of her depression. We were astounded to see how quickly she went from a depression that verged on suicidal to high energy and optimism.

The next step was to conduct controlled clinical trials of dawn simulation. Seattle researcher David Avery pioneered this effort. He was able to show that a dim red, brief "placebo" dawn had less effect than a brighter white, extended dawn. For our follow-up work, we used a halogen lamp encased in a diffuser on a tripod next to the bed, with the light aimed at the pillow. The microprocessor that controlled the brightness gave us the ability to simulate the progress of dawn on a particular day of the year at a particular latitude and longitude. Volunteers were assigned to one of three groups. One group received the dawn simulation; a second got a dim on-off pulse from the lamp that matched the total light output of the dawn signal; and the third did bright light therapy after they woke up. The results indicated that dawn simulation was as effective against depression as bright light therapy.

The technology of clinical dawn-dusk simulation is still undergoing research and development. We expect that the first sophisticated, affordable systems will become available in the next couple of years. It is already possible to buy an electronic controller with flexible dawn-dusk adjustments and hook it up to a large, diffused light source positioned above the bed (see *Resources for Follow-up*, p. 301). The light, from a lamp behind a diffusor screen, floods the entire top of the bed. For now, this is the best way to try this new-generation light therapy at home. Some users have found that it works every bit as well as bright light therapy. Others have gone on using their light box—if only briefly—after they wake up, for an extra morning energy charge.

Meanwhile, the concept of dawn simulation has begun to spread in ways we didn't anticipate. A recent online search for "dawn alarm clock" garnered over two million hits! Dozens of companies have started marketing alarm clocks with built-in lights that get brighter over some preset length of time. Many of these appliances have other features as well, from radios and MP3 players to "scent therapy" and recorded bird songs. The first thing that strikes us about these devices is that they are much too small to flood the bed space with light. That makes it all too easy to roll over while sleeping and miss the signal. And no one, to our knowledge, has conducted the sorts of controlled clinical trials that would prove whether they are effective. If you like the idea of being awakened gradually by a bedside lamp that plays music, smells like spruce or lavender, and trills like a lark, you may enjoy having one of these gadgets. But you should not expect the same results that bright light therapy or clinical dawn simulation offers.

If you decide to undertake bright light therapy or dawn simulation therapy, you need to be aware that you are in for a learning experience. In our controlled clinical trials, participants necessarily followed the same routine, such as 90 minutes of dawn light, rising to 300 lux. Would this particular combination of timing and dosing be exactly the one that is best suited to your individual needs? There is no way of knowing in advance. You may be able to make the necessary adjustments intuitively, but be aware that it is easy to make mistakes self-dosing. Often a clinician's guidance is needed in the beginning.

You will probably face important practical issues, too. What if you need to leave by a certain hour to get to work? How do you get the kids off to school when you're sitting in front of your light box? What do you do on the weekend when you absolutely must sleep in to make up for lost sleep during the week? If you sleep alone you have no one to please but yourself if you want to try out dawn simulation. But lots of people have a spouse or partner who may be less than thrilled to have an artificial sunrise in the bedroom every morning if it comes earlier than they want to wake up.

8. Nighttime Meds and Melatonin

You make a point of getting to bed by 11 PM, but even as you lie down you can tell that this is going to be another one of those nights. What's the matter? Why won't sleep come easily? You have to be alert for that meeting in the morning. Maybe you should take a pill. But what if it doesn't work? Or what if it does work but leaves you so groggy in the morning that the meeting is a loss anyway?

Millions of people face situations of this sort. For some, it happens only now and then, and for others only during short stressful periods. For a great many, however, it happens practically every night. This is *chronic insomnia*, and it is notorious in the medical community for its many possible causes. Among these are anxiety, depression, attention deficit hyperactivity disorder, restless leg syndrome, periodic limb movement disorder, obstructive sleep apnea, and various drug side effects. Then there is the mysterious "primary sleep disorder"—a diagnosis generally given when no other cause turns up.

Very often, however—and far too often overlooked—the source of the insomnia is an issue with the person's inner clock. The clock is doing its job, regulating the daily sleep/wake cycle, but the cycle itself does not fit neatly with the sleep pattern the person wants or needs. Instead, the cycle has become *displaced.* For example, you need to get to sleep by 11 PM, but your circadian rhythm doesn't send the sleep signal until an hour or more later. The result?—an hour of lying awake trying to get to sleep. Such a displacement may also trigger or aggravate the problems linked to poor sleep. If it's severe enough, it can lead to *delayed sleep phase disorder* (DSPD). DSPD victims find it very hard to fall asleep before 2 AM or even later and are at serious risk for depression.

In our view, if you have sleep difficulties, the first thing you should do is check into whether your circadian rhythm is responsible. Clearly, if you have such problems as restless leg syndrome or sleep apnea, these should be dealt with directly. But if the source of your sleep difficulties is less clear, it makes sense to find out if circadian problems are involved before starting a long, arduous course of medication trials.

For example, ask yourself whether you are sleeping—or wanting to sleep—at the wrong hours to fit your daily schedule, if you are fatigued or depressed mainly at certain times of day, or if you have tried sleeping pills but remain awake for more than an hour. Take the online chronotype questionnaire (see *Resources for Follow-Up*, p. 301), which will tell you if you are falling into an extreme group for which we specifically recommend chronotherapy. If your sleep difficulties *are* the result of a problem with your inner clock, this approach offers a straightforward, healthy, superior way to deal with them. Chronotherapy helps with the daytime aftereffects, too—the fatigue, grogginess, irritability, and depressed mood. As these become less worrying, that in turn makes it more likely that you will sleep better, creating a "virtuous circle."

There is a virtuous circle at work for those with DSPD and depression, too. As chronotherapy gradually moves the end of their internal night to an earlier hour, they find that their sleeping pattern smoothly adjusts toward a more normal cycle. At the same time, their depression lifts. Which is causing which? Does the light treatment have a direct antidepressant effect that then resolves the sleep problem? Does getting their circadian rhythm in sync with sleep have an antidepressant effect? Or is it that their improved mood makes sleep come easier, at the same time that improved sleep lifts their mood?

As a rule, these patients are too relieved to much care *how* it happened. However, it is valuable and important to understand how different approaches to insomnia work. If you decide chronotherapy may be best for you, unless your problem is mild you will almost surely need your doctor to get you started, to monitor, and to fine-tune the treatment based on the way you—as an individual—respond. But if you find the treatment helps, in the long term it will most likely be up to you to make continued adjustments as you go along.

Sleeping Pills

The most common treatment for insomnia is sleeping pills (known to the trade as *hypnotics*). Even if you have never tried one of these, you have certainly seen the ubiquitous pitches for them, filled with soft music, lyrical images, and soothing voices offering promises of a restful night's sleep. Sleeping pills are crucial profit centers for Big Pharma, and the billions of dollars they invest in advertising them is money well spent. Doctors constantly have to deal with patients who diagnose themselves and demand a prescription for whatever drug is currently being most advertised. This problem is peculiarly American: The U.S. is one of only two countries in the entire world that allow direct-to-consumer ads for prescription medications. (The other is New Zealand.)

What do sleeping pills do? To work at all, they must be having an effect on the sleep centers of the brain, but just saying that doesn't explain much. There are distinct categories of sleeping pill, which have different neurochemical actions, different levels of effectiveness for different people, and different sets of side effects.

The drugs in the *GABA* group are currently the most widely prescribed. They include the brand names Ambien, Lunesta, and Sonata. Named for their active chemical, gamma-aminobutyric acid, these drugs attach themselves to specialized GABA receptors on neurons that promote sleepiness but might be staying too active at night. Depending on the dose and the rate at which your body breaks them down, these chemicals generally pass out of the system by wake-up time. There is some risk that they remain during the day.

When these drugs are used constantly, their effectiveness against insomnia seems to wane. In fact, after a few weeks of use it is hard to detect any difference between those using them and those using a placebo. More troubling, a very small minority of people react quite badly to them. Their nighttime behavior on the drugs has been as bizarre as driving the car, raiding the fridge, making phone calls, having sex—and not remembering any of it after waking up. It would seem, then, that these pills must activate more parts of the brain than sleep-specific neurons. And if you happen to have both depression and insomnia and take GABA drugs, the depression may actually get worse even if you are falling asleep more easily.

Benzodiazepines are a separate class of sleeping pills that includes Ativan, Halcion, Restoril, Valium, and Xanax. These have been around for many years and are still widely prescribed despite their downsides. Benzodiazepines also work by stimulating GABA receptors, but not specifically the receptors on the sleep-relevant GABA neurons. As a result, they can act to relax muscle tension and relieve anxiety, and are sometimes prescribed specifically for daytime use against anxiety disorders. They can also confuse memory and alter mood states. This gives them a potential for abuse, such as getting a Valium "high." They should never be used for long periods of time, but in fact they often are.

Short-acting benzodiazepines such as Halcion may help you get to sleep, but because they wash out of the system quickly, you may also have a withdrawal-like effect that wakes you up in the middle of the night, for little or no net benefit. Longer-acting benzos such as Restoril may increase sleep by only twenty minutes or so. Compared to GABA drugs, the benzodiazepines carry a greater risk of next-day side effects, including fatigue, inability to concentrate, and a sense of muscle weakness.

A more recent addition to the list is the class of *melatonin agonists*, most prominently Rozerem. These drugs were developed in the lab to mimic the action of the hormone melatonin, which is produced naturally by the body. There is no good reason to believe that melatonin agonists are physiologically more active than melatonin itself, which is available without prescription in the U.S. and costs much less. When melatonin agonists are taken according to the manufacturers' instructions, shortly before bedtime, their effect on sleep onset is small—about twelve minutes on average. Melatonin, too, generally has limited effect on sleep because it is dosed and timed incorrectly, as we discuss below.

Whatever type of sleeping pill you take, if it is working properly you should fall asleep in less than half an hour. If you're still up one or two hours after taking the pill, it's not working, and increasing the dosage is unlikely to make it work. There is another problem, too. As the dosage increases, it becomes more likely that if you decide to quit and try a different approach, you will have trouble getting off the drug. With some short-acting drugs, you may suffer withdrawal symptoms immediately, even the same night. As the drug wears off, you may wake up agitated. If you've been using a sleeping pill for even a couple of weeks—and especially if you've started increasing the dose to keep it working—a night without it can be miserable, far worse than before you started the drug.

Benzodiazepine drugs are notoriously difficult to taper. Your body can sense even the smallest dose reduction, and you probably feel you cannot afford to lose sleep (literally!) over it. So you go back to your earlier, higher dose. Even the newer generation GABA drugs can present a major challenge. It took one of our patients a whole year to reduce the amount of Ambien he was taking to a minimal dose. Even then, he couldn't take that last step and eliminate it entirely without going through withdrawal symptoms.

A Doctor's Critique

One expert who challenges the claims Big Pharma makes for their sleeping pills is Dr. Daniel Kripke of the Scripps Research Institute in La Jolla, California. He writes: "The sleeping pill industry would like you to believe that insomnia leads to depression, implying that sleeping pills might prevent depression. It isn't so. Controlled trials show a higher rate of developing depression among those given the sleeping pills as compared to those given placebo. Perhaps the true mechanism is that insomnia leads to sleeping pill use, which in turn leads to depression. Further, there is extensive evidence that sleeping pills on average impair performance and memory on the following day."

So we are juggling three interrelated factors—insomnia, sleeping pills, and depression—whose direction of influence is hard to tease apart. The old school of psychiatry blamed depression as a trigger for insomnia, and insomnia is still considered a "symptom" of depression. Later thinking blamed insomnia as a trigger for depression, which prompted sleeping pill use. And now Dr. Kripke points to sleeping pills themselves as a trigger for depression. Probably all these analyses have some truth, and the triggering effects are multidirectional.

Dr. Kripke also points to unpublicized statistics showing that mortality itself may rise with sleeping pill use. "Even when people were matched for age, sex, race, and education, and a total of thirty-two health risk factors, those who reported taking sleeping pills thirty or more times per month had 30 percent more mortality than those who said that they took no sleeping pills. The smaller risk of taking sleeping pills just a few times per month was 10 to 15 percent increased mortality."

The research Dr. Kripke cites does not indicate *why* sleeping pill use increases mortality. But does it matter very much that we do not yet understand the chain of events, when the risk is an earlier death?

Melatonin—Hormone and Chronobiotic

What does it take for you to get a good night's sleep? Here is a pocket summary. (For further explanation, take a look at Chapter 2.) There are two separate processes involved in going to sleep. The first is *sleep pressure*, which builds up while you are awake. When it reaches a certain point, you feel sleepy. The second is your *circadian rhythm*. Your inner clock expects sleep to start at a certain time that varies from one person to another. About two hours before that time, the clock sends a signal to the pineal gland to start releasing small quantities of the hormone *melatonin*. This acts as a healthy, physiological signal of darkness for the brain and body. If you are an average sleeper, your melatonin levels start to rise around 9 PM, and you are ready for sleep around 11. The melatonin is not *putting you* to sleep, the way a sleeping pill would, or you would fall asleep sooner. Instead, it is feeding back to the inner clock with a confirming signal that physiological nighttime is beginning. The inner clock then sends a signal to the sleep centers of the brain.

If you are someone who has searched for a way to deal with your sleep problems, you have probably heard of melatonin already. Synthesized melatonin, sold over the counter in the U.S., has become a multimillion-dollar industry. It has also earned a poor reputation among professionals and insomniacs, who say that it simply doesn't work as advertised. If you take it twenty minutes before bedtime, the way it says on the label, you may or may not get to sleep more easily. Even if it does seem to help sometimes, you may wonder if this is only a placebo effect. Would taking anything touted as a sleep aid have worked just as well?

There are at least two reasons to be dubious about using drugstore melatonin as a sleeping pill. Once the pineal gland begins to make its own melatonin, a couple of hours before bedtime, the hormone starts circulating widely throughout the body. Adding a melatonin pill right before sleep is unlikely to do much, since the nightly action is already under way. Furthermore, the usual drugstore dose of 0.5, 1, or 3 milligrams—and even higher—puts tremendously higher levels of hormone into the blood than the pineal gland ever produces. How the nervous system reacts to this unusual event is still poorly understood.

Once an immediate-release melatonin tablet dissolves in the intestines, the hormone is released into the blood, then rapidly washes out of the system. With controlled-release tablets, melatonin remains in circulation until sometime the next day. This raises the problem of morning or even morning-to-midday hangover. Even with immediate-release tablets, the hormone can linger into the next day if taken in megadoses, simply because there is more of it to wash out.

Given these problems and doubts, why do we even raise the subject of melatonin? The reason is this. As a sleeping pill—a medication that pushes the body toward the sleep state—it is mostly a flop. As a "chronobiotic" tool—one that can synchronize circadian rhythms with day and night outdoors—it is first rate. Like bright light therapy, melatonin, properly used, can precisely reset the inner clock. Both methods shift the inner clock earlier, but with an important difference. Melatonin taken in the *evening* shifts the clock earlier, while light exposure in the *morning* shifts the clock earlier. The two methods can be used individually, but they can also be linked for a stronger total effect.

Less Is More

How can melatonin be used to move the inner clock earlier? A key insight was that *lower doses work better than higher doses.* Of course this is just the opposite of the way we ordinarily think about drugs. The secret is to provide tiny levels of melatonin, no higher than the levels the pineal gland releases at night, but to provide them in the late afternoon or early evening, hours before the inner clock signals the pineal gland.

In a pivotal study by Dr. Alfred Lewy at Oregon Health & Science University, research volunteers took a series of 0.1-milligram tablets throughout the late afternoon and evening, starting about six hours before normal sleep onset. (Repetition of this ultra-low dose was necessary to keep the signal active, since the system washes out the substance so quickly.) Thus, hormones bathed the inner clock with pineal-level melatonin for several hours before the pineal gland switched on for the night. The result? The inner clock shifted earlier, so the circadian signal for sleep onset came sooner.

In our lab, we set ourselves a new goal: to make this experimental result practical in a clinical setting. Realistically, we could not expect patients to take microdose tablets repeatedly, on a strictly timed schedule, throughout the evening. This would also create a series of spikes in blood concentration following each tablet, which is quite different from what happens naturally—a gradual but steady release of pineal melatonin.

The challenge was twofold: first, *timing* the release of melatonin to match the pattern of pineal production, and second, *limiting* the level of melatonin that reached the blood to the pineal range, which is much lower than in existing commercial products. To develop a controlled-release tablet, we micronized chunks of melatonin, reducing them to particles no larger than 9 microns, which is about the width of a single strand of spiderweb silk. This ultrafine powder was then mixed with supporting particles that dissolve slowly in the digestive tract, allowing a fine stream of melatonin to be absorbed over hours.

The next step was to find out if our tablet really did mimic the way the pineal gland releases melatonin. Research volunteers agreed to have blood drawn repeatedly from early evening until noon the next day. Because the pattern of pineal melatonin release is different from one person to the next, we carried out three variations on different nights. On one, the volunteers got a placebo tablet that did not contain any melatonin, to establish their natural pattern. On another night, we gave them our new controlled-release tablet. And on a third night, they took a higher-dose tablet like those that are sold commercially.

The results were everything we had hoped for. The microdose tablet produced levels of melatonin in the blood in the same range as that produced by the pineal gland of a healthy young adult. Even more gratifying, the melatonin from the tablet washed out of the blood toward morning, exactly the way pineal melatonin does. In sharp contrast, the drugstore dose was still lingering when we stopped measuring at noon the next day. It was clear that we had a potential clinical tool to deliver a circadian signal to the master clock.

We have successfully tested this "naturalistic" melatonin formulation with patients whose circadian insomnia ranges from mild difficulties at bedtime to extreme delayed sleep phase disorder (DSPD) with sleep onset as late as 8 AM. (See *Resources for Follow-up*, p. 301.) As a rule, the patient takes the tablet six hours before the desired sleep time, or about four hours before spontaneous pineal melatonin production begins in response to signaling by the inner clock. Because of the controlled-release timing, the tablet melatonin washes out by the middle of the night, so the inner clock does not get a conflicting instruction to shift later.

Cliff M., twenty-eight, was one of those with extreme DSPD. He could not fall asleep before 7 AM and habitually slept until 3 PM. Sleeping pills didn't work for him, and alcohol only got him to sleep a couple of hours sooner. A successful scientist, he pulled all-nighters to keep up in his job, but he missed out on interactions with his coworkers. We had him take the microdose, timed-release melatonin six hours before his usual sleep time, while he was still wide-awake, and wear blue-blocking wraparound glasses until bedtime to eliminate the antagonistic action of such light exposure.

The blue-blockers served three purposes. First, blue light signals the retina and the inner clock that it is daytime, while melatonin signals that it is night. Their mixture is a formula for physiological confusion. Second, a melatonin tablet in the evening is designed to shift circadian rhythms earlier to enable earlier onset of natural sleep, while blue light in the evening and most of the night shifts the rhythms later. Finally, melatonin bathes the retina to enhance dim light perception at night, and blue light during this period may cause a photosensitizing reaction with damaging consequences for the eyes, especially with repeated exposure.

In addition to his pre-sleep routine, Cliff also did bright light therapy the next day as soon as he woke up. The light box produced a broad, white spectrum that included the blue that was restricted at night, providing the physiological signal for the start of the day. His inner clock immediately started edging earlier, step by step. After only two weeks, he found it easy to sleep from 11:30 PM to 7 AM. "I couldn't believe it," he said. "I was sure my clock was permanently damaged!"

What if you are someone with mild circadian insomnia, who cannot fall asleep before 1 AM or so? Not that unusual a situation, certainly, but it may leave you faced with an unpleasant choice between getting too little sleep and being habitually late for work. Our advice to those in this predicament is to take the chronotype questionnaire (see *Resources for Follow-up*, p. 301). This provides an instant estimate of pineal melatonin onset time. Count back four hours, and that's when to take the controlled-release microdose melatonin. It will be about five hours before you reach bedtime, though.

It is important not to let evening light exposure work against the effect of the evening melatonin. This doesn't mean sitting in the dark, however. Low-level room light from incandescent or compact fluorescent bulbs with color temperatures of 2,700–3,000 kelvin is okay. Using the computer is okay, too, if the screen color is adjusted using the f.lux application (see *Resources for Follow-up*, page 301). For an explanation of color variations, take a look at Chapter 5. If you must go out into bright light after taking evening melatonin, you can use blue-blocking glasses to avoid interfering with the melatonin's effects. In contrast to sunglasses, these will enhance visibility in the evening and night.

A Boost for Weak Rhythms

We generally think of circadian insomnia as a problem for owls whose chronotype conflicts with the sleep/wake cycle they need or want to follow. But not necessarily. Sometimes a sleep problem results from *weak* rhythmicity, a circadian system that is not quite up to the task of controlling the sleep/wake cycle. We have found microdose melatonin to be helpful in cases of this sort, too.

Melissa N., sixty-six, was self-employed and had flexible work. Her chronotype was neither morning nor evening, and her sleeping pattern was unpredictable from one day to another. She found it easy to stay up very late if she was involved in reading or watching a movie, which happened a few nights a week. When she needed to, because of an early appointment the next day, she had no trouble getting to bed early. Sometimes these two patterns came into conflict, however. She might get involved in something late in the evening, get to sleep quite late, then have to wake up early the next morning after only four hours of sleep.

We were not sure if this was a matter of circadian insomnia or simply a lack of self-control with poor sleep hygiene, but Melissa had a positive experience with melatonin. On evenings she got involved in activities but knew she would need to wake up early, she began to take a microdose after dinner. This eased her into sleep before midnight, when otherwise she would have chosen to stay awake. Her use of melatonin was discretionary—not every night—so she could continue to enjoy her variable schedule at will.

Circadian Insomnia and Sleeping Pills

What if the underlying issue that leads to insomnia is circadian rhythm displacement—can sleeping pills play a useful role? Based on our experience, no. For someone with circadian insomnia, the sleep offered by a sleeping pill is *false* sleep, out of sync with the person's inner clock. Yes, sleep may shift earlier, but there is still that mismatch between the external day/night cycle and the internal sleep/wake cycle. Worse, forcing sleep earlier in this way quickly depletes the sleep pressure that has been building all day. When the circadian signal does arrive, perhaps hours later, it does not have the advantage of being in synchrony with high sleep pressure. This sort of sleep is often unrefreshing and less than helpful for waking activities the next day.

It can also happen that someone with circadian insomnia takes a sleeping pill at, say, 10:30 PM but is still awake at 1 AM. Waking up for work becomes daily agony. On weekends, sleep onset easily drifts to 2:30 AM, followed by sleeping in for several hours in the morning. The drug is supposed to sedate quickly, but it doesn't, and switching drugs does not help. In these cases, the inner clock is winning out over a delayed, weak hypnotic action of the drug. The combined activity of the inner clock and the drug may generate a compromise, with sleep becoming possible somewhat earlier, but it doesn't achieve a normal schedule, and circadian insomnia persists.

Someone else may succumb quickly to the sedative effect of the sleeping pill, only to have the effect wear off *before* the inner clock is ready to signal sleep. Georgina P. was a fifty-two-year-old home-office worker whose husband went to sleep regularly at 10:30 PM. She felt a strong obligation to go to bed at the same time as her husband, but she couldn't do it without taking a sleeping pill. Then at 2 AM, she would wake up fully alert. She would not be able to get back to sleep until an hour later. If you added up her total sleep time, it was normal. But she was plagued by this "middle insomnia" and felt groggy for several hours after forcing herself awake at 7 AM with the help of an alarm clock.

The clue to understanding her problem was in her chronotype questionnaire score. It was practically at the bottom of the scale, indicating extreme owlishness. The sleeping pill knocked her out at 10:30 PM, but its effect wore off after about three hours. Meanwhile, her circadian clock was set for a 3 AM sleep onset. As a result, she would be awake from 2 to 3 AM before starting her natural sleep episode. And when she forced herself up at 7 AM with her husband, her circadian clock was still in its nighttime mode, causing the morning grogginess.

The first step in Georgina's chronotherapy was a frank discussion about the moral obligation she felt to synchronize her sleep with her husband's. "Don't force yourself to sleep so early," we told her. "Wait until the middle of the night when you know the urge will come. Then we can begin to shift your internal clock earlier with chronotherapy." She was reluctant but agreed to try. She put away the sleeping pills and took a microdose of melatonin at 9 PM. Within three days, she reported feeling wonderful for the first time in years. It was such a relief to stay awake late at night, "following my rhythm." And instead of 3 AM, she was getting to sleep at 2 AM. A week later, we moved the melatonin from 9 to 8 PM, and later to 7 PM. After about three weeks, she was able to get to sleep by midnight, just an hour and a half after her husband, with no more sleep interruptions in the middle of the night. Greatly relieved, she decided to continue on that schedule.

Getting Off Sleeping Pills as You Transition to Chronotherapy

Sleeping pills may "work" merely because of suggestion—the placebo effect, which can be long-lasting. Or they may work by acting on GABA receptors in the brain's sleep centers. Whatever the reason—which is impossible to ascertain for any individual patient—the more you use them, the harder it is to quit even if they are no longer working. Sleeping pills may also have become addicting above and beyond their sedating effect. We see this in patients who need higher and higher doses and experience withdrawal symptoms if they try to skip or reduce the dose. That's the reason these drugs are "controlled substances," with strict limitations on prescribing.

When we encounter this problem with patients who are turning to chronotherapy, we really cannot insist that they stop taking the sleeping pills before starting melatonin, or melatonin plus morning light therapy. Initially we don't even try to change the dosage. Instead, we ask them to carefully monitor themselves, to watch for new sensations of sleepiness that come *before* they have taken a sleeping pill. These indicate that the inner clock is starting to shift earlier. As soon as they feel them, they should go to bed. Even then, they may continue for a while to take the drug as they get into bed, but *not* beforehand. Once the circadian rhythm synchronizes with sleep, we can start to taper off the sleeping pill.

Some patients manage to withdraw from sleeping pills very quickly. Dr. George L. is a thirty-three-year-old hospital physician who works ridiculously long twelve-hour shifts, starting at 6:30 AM. He was a strong evening chronotype who found it easy to stay awake until 2 AM when he could, but his work schedule made that impossible. The only way he could get what he considered an adequate night's sleep was by taking Ambien at 10 PM, which put him to sleep by 10:30 PM. He began chronotherapy by taking microdose melatonin at 8 PM—six hours before what we estimated would be his drug-free bedtime. For the first three days, he took the Ambien as usual, at 10 PM. Then he skipped a day and discovered he could still get to sleep. Nervous about whether that would last, he resumed the Ambien for another couple of days. Then he made up his mind to quit cold turkey. No tapering off, no withdrawal symptoms—and he continued successfully with melatonin alone.

How Long to Go?

One concern most patients have is how long their chronotherapy will have to continue. It requires daily attention, after all, and has to compete with their other priorities. Well, if you have classic delayed-type circadian insomnia, the pressure to slip later is a basic aspect of your inner clock. Chronotherapy does not change that—it keeps it under control. If you stop treatment, you are likely to drift back into a later sleep pattern. However, once you've met your goal, there are other steps you can take. You can try going outside soon after you wake up. Take a walk, jog, or bike ride—during the months when the sun is already up, of course—to keep your clock in check. If you are using both melatonin and light therapy, you can test whether evening melatonin alone will do the job. The details will differ from one person to another. But even if you slip, you will already know how to start over with a new course of chronotherapy. You may have done it under a doctor's guidance the first time, but if you need to do it again, you can probably do it on your own. Still, we have to face the fact that some people will want or need to keep up their regimen indefinitely. We know some who have been doing chronotherapy for twenty years, and they are the first to attest to how much it has improved their quality of life.

9. Hospitalized with Depression

Imagine that an understanding of chronobiology led to the development of a new treatment for the most widespread, severe manifestation of psychiatric illness. Suppose that this approach produced results much sooner than the drug therapies most commonly used. And what if, in addition, this approach had negligible side effects, required no expensive technology, and called for no additional hospital staff or complex retraining? Wouldn't the psychiatric world rush to put it in operation?

Surprisingly, a treatment method with these qualities *does* exist. It can produce improvements in patients in as little as a single day. This is easily the fastest turnaround known to psychiatry. But even more surprising, this dramatic breakthrough is only beginning to overcome skepticism and find acceptance in the medical community. The treatment is called *triple chronotherapy*, and the challenge it meets is the resolution of a major depressive episode.

The Priority in Emergencies: Medical Intervention

Depressive disorders are far and away the most common serious psychological problem. The statistics are staggering. In a given year, about one in ten Americans—9.5 percent of the population—is suffering from a clinically significant depression. That adds up to about 18.8 million people, with almost twice as many women as men (12.4 *vs.* 6.4 million).

People with depression are heavily represented among those who are hospitalized. In 2005, nearly one in ten of those admitted to hospitals in the U.S.—about 2.9 million patients—were suffering from a depressive disorder. Many of these entered the hospital with other problems as well, such as cardiovascular and respiratory conditions, but for almost half a million of them, the depression was the major reason they were there.

Why does someone with depression show up at the hospital and get admitted as an inpatient? One major reason is that the person has begun thinking seriously about suicide, perhaps to the point of making plans, talking about them to others, and giving away treasured possessions. This can become so disturbing to friends and family members, as well as to the individuals themselves, that they are driven to ask for help, often at the psych emergency room. Some may have actually attempted suicide. In those cases, even if they try to reject the idea of getting medical help, their families, and society as a whole, insist on it.

If you develop feelings about death that are vague but disturbing, it is important to share them with someone close or to speak confidentially with volunteer experts who can guide and assist you (check befrienders.org, which offers contacts worldwide). If these feelings become so strong that you are uncontrollably focusing on your own demise, you or someone you trust needs to take immediate action. Go to the nearest emergency room or dial your local emergency number (911 in the U.S.). Help is there to calm these intense thoughts, but it depends on you to seek it.

Another common reason for hospitalization is that someone being treated for a depressive disorder has had a bad response to the medications their psychiatrist or primary care physician has been prescribing to them on an outpatient basis. Perhaps the person is experiencing severe insomnia, heightened agitation, or an allergic reaction. A bipolar patient may have broken through the depression into a sudden, dangerous mania. Or the depression may have worsened to the point that a whole new treatment strategy needs to be administered under twenty-four-hour observation. If so, a common starting point is to discontinue and wash out the current drugs completely, then embark on a new approach to medication.

What if the new drug strategy fails to give good results, even under such close supervision? At that point, patients are likely to be given a series of electroconvulsive therapy (ECT) sessions every few days, while the drugs are maintained.

Part of the reason for bringing in ECT so soon is that its antidepressant effect kicks in relatively quickly. In contrast, medications can require weeks to take effect. This leaves the doctors wondering for a month or more whether the particular drugs they have prescribed are working or not. If it turns out that they aren't, still another treatment strategy will have to be attempted.

These days, rapid improvement is very high on the list of clinical priorities. Psychiatric hospitalization is expensive, as much as several thousand dollars a day, and insurance companies often put pressure on the hospital to discharge patients before it is therapeutically wise. When a patient is forced to leave the inpatient unit too soon, the mood of the staff veers between anger and despair over the way our systems for medical coverage do not consistently place top priority on the patient's needs. Without the continuous attention from expert staff, the patient's chances for improvement are markedly lower. Said one dejected colleague, "It seems as if our job is more one of managing drug side effects than getting our patients better."

The situation for the patients we have just described is both dire and self-perpetuating. Huge numbers are hospitalized with depression every year, treated with powerful medications and electroconvulsive therapy, then often sent home before their condition has been adequately treated. The psychological and economic costs are enormous for them, their families, and society as a whole.

Larry's Story—Part I

Larry N., twenty-one, is a college student who grew up in the Chicago area and is currently on leave from the University of California at Berkeley. "I really want to be an architect," he says. "I don't care how long it takes. I don't know if Berkeley is good for me—I have a lot of friends out there, but there is a lot of partying."

In the middle of his freshman year, Larry went through a major manic episode. "I had been home for the winter break," he recalls. "Then I flew to the Caribbean, then to Chicago, and then back to California. So I crossed three time zones and I had a couple of beers, and it started from there. I was sleeping maybe an hour a day. My friends brought me into the ER. And then I went up to the psychiatric floor in the hospital. I remember random spurts, but things were moving so fast I can't remember much.

"I was diagnosed bipolar," he continues. "I was in the hospital at the university for a few days. Then I came back to Chicago, and my soon-to-be sister-in-law said she knew a specialist. That's how I got set up with Dr. [John] Gottlieb [at Chicago Psychiatry Associates]. I was twenty years old."

Larry felt able to go back to college for the spring semester. "It seemed to go well. I did well and I had a good summer. So I went for the fall term. That was when I had just turned twenty-one, and I got into the engineering program for the winter term. I was really happy about that. I had a rigorous schedule. Because the school tries to weed out people who aren't going to make it, you have to do well.

"I was having a good time. I met people and I had friends in my studio. I was also playing rugby. *Not* lacrosse—rugby," he adds with a laugh. "People get them mixed up."

But soon the pressure began to affect him. "I wanted to do engineering so bad, I knew I had to shape up and get up for class. But something didn't work out right. Because I was drinking, I started pushing my sleep back, and then I was missing class. I got sick. I couldn't get out of bed. I couldn't concentrate or do any of my work. I got really depressed. I came home in April and I started seeing Dr. G again."

Dr. Gottlieb recommended a course of triple chronotherapy—three days of wake therapy followed by phase-advanced recovery sleep and bright light therapy.

"He knew my whole story," Larry says, "and said I had three options. I could take Wellbutrin for six months and we could see how that worked. Or we could try Lamictal [a mood stabilizer, more than a direct antidepressant]. That could work, but some people are allergic to it. Or we could try the chronotherapy, which could work faster. He thought I was perfect for it. I wanted to not be depressed anymore, so I did it."

Sleepless Nights

Despite its advantages, chronotherapy is just beginning to be used widely as a treatment for depression. Why has the medical community been so reluctant? The reasons are as much historical as anything else. The origins of this treatment go back some forty years, to a hospital psychiatric ward in Tübingen, Germany. A woman who was hospitalized for depression kept saying that she could relieve her own symptoms by taking all-night cycling trips. It didn't sound very likely, but a young doctor named Burkhard Pflug was so struck by her insistence that he decided to look into her claim. Testing hundreds of cases over the next decade, he found that when depressed patients were kept awake all night, a majority of them said they felt wonderful the next day, even if they had been in the extreme depths of depression only a day earlier!

This reversal in clinical state was startling and mystifying, but it showed up again and again. Most of us would probably expect that staying awake all night would be like suffering from an extreme bout of insomnia that would lead to lethargy, disorientation, worsening depression, and uncontrollable sleepiness the next day. But no, just the opposite. Patients became cheerful, even elated. They started paying attention to their personal hygiene and appearance, and chatted volubly with staff. And of course, many said they wanted to go home immediately, now that they were miraculously "cured."

The story didn't end there, however. After sleeping the next night, many patients crashed back to where they had been before the night of not sleeping. The recovery effect was dramatic, yes, but for most it was also transitory. Patients obviously could not be kept awake every night to maintain the benefit. So instead of becoming the newest medical weapon against depression, sleep deprivation took on the status of a curiosity. It might be mentioned in a footnote, but the slower, weeks-long use of drugs remained the mainstay intervention. This was so even though patients who went through a full course of medication rarely ended up feeling as well as they had after a single night of sleeplessness.

In the mid-1990s the story began to change radically. A few research psychiatrists—notably Francesco Benedetti in Italy, Alexander Neumeister in Austria, and Joseph Wu in California—were fascinated that a single sleepless night could give such immediate relief to someone in a deep depression. What if a way were found to prolong the effect and provide sustained relief, without the impossible step of banishing sleep? The potential benefits were immense, both to patients themselves and to overburdened, underfinanced psychiatric facilities.

As they soon realized, a large part of the answer lies in making adjustments to the circadian timing system. It turns out that not all sleeplessness is equal. Staying awake an entire night has an immediate and powerful effect on depression, but staying awake only through the *second* half of the usual sleep period can also have an impact. The reason, apparently, is that this unusual wake period, even under ordinary room lights, helps to advance the circadian clock. We have also seen in earlier chapters that exposure to bright light somewhat before the person's usual wake-up time has a similar effect on the circadian clock, but without the sleep deprivation that triggers the sudden improvement.

There had already been some strong hints of this effect. At the hospital in Milan where Benedetti practices, researchers noticed that patients with bipolar disorder who were assigned to rooms with east-facing windows left the hospital significantly sooner than those in west-facing rooms. Apparently these patients were receiving beneficial morning doses of light therapy without anyone intending or even realizing it at the time. Since then, controlled studies have confirmed this informal observation.

The three critical elements were now in place for what has come to be called triple chronotherapy:

- *Wake Therapy.* Patients stay awake through the night and into the next morning. Once their normal day begins, the circadian clock issues its usual wake signal. This keeps them up through the day, in spite of accumulating pressure to sleep.
- *Recovery Sleep with Phase Advance.* Toward the end of the day, but several hours *before* their usual sleep time, they are encouraged to sleep for eight hours. While this recovery sleep meets their sleep needs, physiologically they are still in the middle of the wake phase of their particular circadian cycle.
- *Light Therapy.* Each day of the treatment, an hour earlier than their ordinary wake-up time, they spend a half hour at a bright light box.

Working together, these elements lift the patient's depression and extend the relief beyond that first, astonishing day.

For many people, one sequence of triple chronotherapy produces such improvement that after a few more days of observation, they can go home. Others may need a second or even a third sequence, but even then, the course of treatment typically takes little more than a week. That is quite a difference from the four to six or more weeks needed to treat the same condition with iffy medications. Once home, they can continue daily light therapy to maintain the improvement.

The Hospital Protocol

Let's imagine that you could track a patient with bipolar disorder who is in a deep depression and has just been admitted to the hospital. The medical team decides that our patient—call her Carol—is a good candidate for triple chronotherapy. She is given a mood-stabilizing drug such as lithium, to protect against bipolar switching to mania, and her other medications are discontinued or adjusted. She is also urged to stay with her typical sleep schedule when she is at home.

DAY ONE: If Carol is like a great many people with bipolar depression, she will stay awake until quite late, well after lights-out on the ward. Let's say she finally falls asleep at 1 AM. Once that happens, she is left alone. No nurse comes in during the night, turns on the lights, and measures her vital signs. Instead, she is allowed to go on sleeping until she wakes up spontaneously. The staff carefully notes her wakeup time (10 AM) and gives her breakfast, even though the other patients have long since finished eating.

Carol's first day in the hospital, with all the stresses of intake procedures, separation from family, and adapting to a new environment, may have disrupted her spontaneous sleep cycle. If that seems to be the case, nothing much happens on the second day. Once again she is urged to maintain her usual sleep pattern, which the staff records. The active treatment will wait for the next day.

DAY TWO: If Carol seems to have handled the admission process well, then today marks the beginning of her wake therapy. No napping allowed during the day, of course, and come nightfall, she must stay awake under normal room lighting. The staff have instructions to check on her several times an hour, to chat with her, to give her lots of encouragement to stay awake, and to offer snacks, coffee, tea, and caffeine-rich candy whenever she likes. She may also have chosen to take a medication that helps her stay awake.

Carol is free to use the exercise room and the computer room during the night. The staff helps keep her alert and watches for any tendency to nod off. The hospital in Milan even offers a garden for strolls outdoors and a fully equipped kitchen where the patient can prepare elaborate gourmet meals to share with other patients the next day.

Remember, though, that Carol came to the hospital in a deep depression. Although she may desperately long for a treatment that helps her, around three or four in the morning—the time that F. Scott Fitzgerald called "the dark night of the soul"—her determination to keep on with it may start to flag. This is the point at which the motivational efforts of the staff are going to matter the most. Following the example of spectators along the course of a marathon, they cheer her on.

"Come on," one aide says. "You can do it! Look how far you've come already. You're most of the way there!"

"This treatment is the whole reason you're here," another reminds her. "In just a few hours, you'll probably be feeling a lot better about things. Isn't that worth the effort of keeping awake a little longer? How about splashing some cold water on your face? Would you like another cup of coffee? Something to eat? How about a walk around the halls?"

DAY THREE: With the help of the staff's encouragement, Carol makes the effort and stays awake. As dawn comes up and she begins her third day in the hospital, she is surprised to find that her sleepiness starts to fade. The chances are excellent that her mood is beginning to lift as well. At 9 AM—an hour earlier than her usual wake-up time—she goes to the light therapy room and spends the next thirty to forty-five minutes seated in front of a bank of bright but not glaring lights. She doesn't peer into the lights, but rather has breakfast and concentrates on reading under the lights. These light therapy sessions will continue, at this time of morning, for the rest of the treatment and after she returns home. They, too, contribute to lifting her mood.

Carol has now been awake for more than twenty-four hours. The pressure to sleep is building in her, but because her internal clock sees this as the middle of her usually alert daytime period, she stays awake. And because her severe depression has been at least temporarily relieved, she is probably active, talking with staff and other patients, reading, watching TV, dealing with e-mail. She may even feel that she is cured and ask to be sent home. If so, the medical staff will explain that she is only partway through the treatment and that it is critically important to complete it. The staff will also monitor her closely, to be sure that she does not nod off during the day.

Early in the evening, five hours before her habitual bedtime, Carol goes to bed in a quiet, darkened room. She is given a sleep mask and, if she doesn't manage to fall asleep within forty-five minutes and asks for it, a low dose of Ambien. As on Day One, her sleep is not interrupted by ordinary hospital routines.

DAY FOUR: Five hours before Carol's normal 10 AM wake-up time—in her case, at 5 AM—she is awakened by a staff member. The room lights go on and she gets breakfast, along with a lot of encouragement to wake up fully and start the day. In this way, just as during the night before, she is awake during the second half of her internal night. But because she has had a full, if early, night's sleep, she finds it easier to handle. The light therapy session, once again at 9 AM, also helps.

There is a good chance that by this time Carol is feeling fine. There is also a good chance that she can maintain this benefit by going to sleep at midnight, an hour earlier than had been usual for her, and waking up an hour earlier for light therapy. If this improvement continues for a few more days in the hospital, she will be ready to go home—with a light box, of course.

If she was started on an antidepressant drug, she will probably continue to take it for several weeks until it takes full effect. But it is also possible that at this point, she may want to discuss with her doctor reducing or stopping the medication, while maintaining the daily light therapy at the earlier time she settled into in the hospital.

What if, on Day Four, Carol is feeling only a little better, or no better at all? In that case, her doctors will advise her to go through a second cycle of wake therapy followed by early recovery sleep. This time the recovery sleep will begin three hours before her usual bedtime, instead of five, but otherwise the procedure is the same. And if this second cycle gives some additional relief but not quite enough, a third wake/recovery sleep cycle may be needed. Even then, however, the entire treatment requires only about a week, plus a few additional days of observation.

We should be careful to keep in mind that triple chronotherapy offers *relief* for depression, not a *cure*. If Carol is typical, once she is home she will need to continue daily light therapy. She may also be on medication, if she and her care provider find it is helpful. But as long as she remains free of the ravages of depression, Carol will probably not worry very much over the difference between relief and cure.

Larry's Story—Part II

On a Friday evening, Larry went to the chronotherapy center, in a downtown hotel. "I got there at 10 PM, and you have to stay up until 6 PM the next day. It was rough. It was really hard to do. I drank coffee, and I had cigarette breaks, and I was splashing water on my face. There was a nurse who stayed up with me first, and she was very nice. Then there was a psych graduate student, then there was another nurse, then another psych student again, and then a nurse. I was getting anxious around 7 AM on the change of shift. I got scared that I would be manic again.

"The next day—that's Saturday—you go to bed at 6 PM," he continues. "You have to wear blue-blocker glasses for two hours before you go to sleep. Then I was awakened at 1 AM Sunday morning and I [had to stay up] until 8 PM Sunday night, and then I slept till 3 AM. Then I had to be up from 3 AM till 10 PM Monday."

Larry responded to the treatment right away. "It was cool, the way it shocked my system. It started working a *lot* faster than I thought it would! Dr. G had told me its value, but I was surprised. I realized that I was feeling better in twenty-four hours, that I was less depressed. When it was time to leave, I was surprised that I would be going home."

Since the wake therapy, Larry has continued with the light therapy. "I know it's a big deal about wake-up time, and I always wake up at the same time and I try to be very good about that," he says. "I try to do the lights every morning (except sometimes). Sometimes I don't. If I haven't done it for a few days, I'm a little more sluggish. But if I'm late to work in the morning, I only do ten minutes instead of half an hour, but ten minutes helps."

What would Larry say to other people about his experience with triple chronotherapy? "You know," he replies, "I'm not trying to make it sound good. It really did work!"

How Chronotherapy Affects Depression

Triple chronotherapy is just beginning to see widespread use, and follow-up research is still scarce. Studies have followed patients for six to nine months and found that the benefits continued without further hospitalization. This compares quite favorably with the odds of a relapse among patients who received standard antidepressant treatments. For those who continue with light therapy and possibly medications to maintain their improved state, theoretically there is no reason that they should ever need to repeat the experience. However, relapse into bipolar depression remains a risk throughout the lifetime, and if and when that time comes, there is no reason not to repeat triple chronotherapy.

But what is it about triple chronotherapy that can lead a person to emerge so quickly from a severe depression? What are the mechanisms involved? A crucial question, certainly, but the most accurate answer is that no one knows for sure.

The same answer, we should point out, holds for antidepressant drugs, too. They can work better than placebos, though placebo responses are common in depression. Not every clinical trial shows an "approved" drug superior to the sugar pill alternative, and not every patient responds to a given drug—which leads us to conclude that all depressions are not equal, but have a wide range of physiological and psychological causes.

If an antidepressant does work, the factors that allow it to work are still a mystery. There are several strong research hypotheses—but as scientists are the first to admit, hypotheses are formed with the expectation that they can or will be disconfirmed.

How can it be that the roots of such a widespread, severe, and widely treated condition as depression are still rather opaque to medical science? One explanation is that medical research often follows a different path than we might expect. The strongest scientific method implies that the logical steps toward discovery of an effective antidepressant treatment are to:

- Define the problem.
- Find out what causes the problem.

- Look for something—a drug, a procedure—that is designed specifically to counteract the cause of the problem.

However, most of the time in psychiatry, the process looks more like this:

- Define the problem.
- Hit by chance on a drug that has been useful treating other illnesses—and notice, by serendipity, that it seems to improve depressed mood. Play with the chemistry to try to enhance the positive effect.
- Assume that brain processes activated by the drug were deficient—the cause of the problem.

In other words, understanding the *cause* of the problem becomes a by-product of the main effort, which is to *correct* the problem.

This situation isn't unique to psychiatry. The history of medicine is filled with treatments that were effective but mysterious. Two thousand years ago, in both ancient Greece and ancient China, people with inflammation or fever were told they should chew on willow bark. No one knew that willow bark contains salicin, a close relative of aspirin. But they did know that it helped reduce fever and inflammation.

Aspirin itself was a mystery cure well into the twentieth century. Purified from willow-bark extract and marketed by Bayer in the late 1890s, it was not until some seventy-five years later that scientists discovered *how* it works.

On the one hand, we don't *need* to know the mechanism of action of a drug before we use it effectively against some illness. On the other hand, the fact that it's effective does not necessarily tell us about the source of the illness. Over the years, researchers have discovered a good deal about how various antidepressants affect the brain, although there is still much to learn. One very important effect is to alter the levels in the brain of neurotransmitters such as serotonin, dopamine, and norepinephrine. Another is to target particular brain structures such as the anterior cingulate cortex, which is thought to play a role in responding to uncertainty and negative events. Do the body's stress responses also play an important role, by changing the way the endocrine system functions? Any—or all—of these factors may help explain depression.

What is significant here is that triple chronotherapy has been found to have the very same physiological effects as medications, but in a more tightly focused fashion. By working in hours (in contrast to weeks, for antidepressants), it could indeed serve as an inspiration for better-informed, future drug design. And beyond its druglike effect, it targets the circadian clock. Perhaps that's the reason chronotherapy works so quickly—by getting to the heart of the matter. The explanatory power has been great, with circadian rhythm maladjustment identified as a factor underlying the whole spectrum of depressive illness—seasonal and nonseasonal, unipolar and bipolar.

Novelty of the Approach—and Obstacles

The benefits of triple chronotherapy are obvious and great, but what about the costs? Like anything new, triple chronotherapy involves some adaptations. It calls for some changes in hospital routine and places some additional responsibilities on the staff. But these adaptations are far fewer, less intrusive, and less expensive than those that almost any other new medical technology would impose. Specifically, setting up a triple chronotherapy inpatient treatment program would mean:

- Reserving some space in the hospital unit or in an off-site facility where patients can sleep or stay awake at unusual times of day.

- Providing, or expanding, facilities such as libraries, computer rooms, and exercise rooms, so that those doing wake therapy have something to do during those long nighttime hours.
- Serving meals at non-standard, individualized times that fit the unusual schedules of the patients.
- Training the staff, especially the night staff, to promote the wake therapy with conversation, coaching, encouragement, offers of snacks and caffeinated drinks, and careful vigilance to make sure the patient stays awake.

There is at least one other significant cost of setting up a triple chronotherapy treatment program. At this point, health insurance companies have not yet given this approach the all-important "procedure code" that allows for reimbursement. Patients who need and want the treatment may have to bear the entire cost themselves. In contrast, Western European health plans generally cover the costs of triple chronotherapy as a treatment for depression, if that is the approach the doctor chooses to follow.

If hospitals continue to drag their feet about offering triple chronotherapy, there is a possible alternative. In 2010, the first outpatient clinic for triple chronotherapy opened in Chicago under Dr. Gottlieb. Located in a corporate housing building, the clinic has access to a computer lounge, private library, club room with TV, full athletic facilities, a kitchen area, and a coffee lounge, as well as an outdoor deck during the summer months. Patients check in for a three-day cycle of wake therapy followed by recovery sleep with phase advance. The recovery sleep takes place in nearby apartments or hotel rooms. The whole experience is supervised and assisted by a staff of psychiatrists, psychiatric nurses, and clinical psychology graduate students with special training in chronotherapy and biological rhythms.

This sort of intensive private treatment is not inexpensive, of course.

Even if insurance companies took a more long-range view of its value, it would still be out of reach for many people who would benefit from it. However, as hospitals and the health care industry realize that triple chronotherapy is cost-effective and can produce quick, complete remissions from deep depression, we envision that they will make it accessible to all who need it. We hope our book—and our readers—will spur the dialogue.

Simplicity *vs.* Supervision

You know the car commercial where the latest model swoops along a curvy mountain road and comes to a screeching halt just inches from a cliff? It ends with a one-second screen overlay: *"Professional driver. Do not try this on your own."*

Triple chronotherapy comes with a similar warning, for a number of reasons.

When people first hear about triple chronotherapy, it sounds so simple—almost unbelievably simple. Feeling down in the dumps? Stay up all night and all the next day. Go to sleep a few hours earlier than your usual bedtime. Sit in front of a light box the next morning, earlier than your usual wake time, and continue the light therapy daily. *Presto!* You'll feel better than you can remember.

Can it really be that easy?

Our answer, as you must suspect, is no. Triple chronotherapy is a medical procedure with a detailed, technical list of issues, cautions, side effects, and counterindications (medical circumstances in which it should not be used). And as someone's psychological condition becomes more severe, the list grows longer and even more cautious. For example, among patients with bipolar depression, about one in twenty responds to chronotherapy by switching to a potentially dangerous manic state. (Those who are put on antidepressant medications run about the same risk, so this is not specific to chronotherapy.) Without monitoring and immediate care by a clinician, such a switch can pose physical, social, and psychological risks that take professional skill to control.

Another important factor is chemical. Many people who have been dealing with depression are already taking a mix of prescription drugs and over-the-counter drugs or supplements, whether prescribed or self-chosen. With some of these, such as Seroquel—which is often prescribed for patients with depression—wake therapy is specifically counterindicated. With others, we don't yet know how wake therapy or light therapy might interact with a particular mix of medications. If a self-treater experiences intensified depressive symptoms—or side effects such as nausea, headache, and agitation—there is no one on hand at home to manage the adverse events.

Still, even with these warnings taken into account, we have indeed seen some people capably manage wake therapy on their own. One patient—a psychiatrist herself—learned that by staying up all night at the end of her work week, she could keep her depression and fatigue at bay until Thursday of the following week, when she usually fell into a slump and needed a repeat dose. But this was an unusual case—someone with the training and skills to monitor her own condition and an observant husband who could make a quick phone call, if needed. Most of us do not have those knowledgeable resources. Nor do we have the information and psychological distance to know just how severe our own condition might be or how much worse it might get.

This much is very clear: People should not try triple chronotherapy for the first time on their own. If you have tried it successfully before under clinical observation and a repeat booster seems to be needed, observe some essential precautions. Consult with your doctor before trying it on your own. Always have someone at hand who can be aroused to help. And make sure you have access to emergency coverage if you should need it.

10.
Beyond Light: The Charge in the Air

Most of us lead lives that rob our brains and bodies of what they need, and we pay a heavy price. As biological creatures, we are exquisitely tuned by our evolutionary history to expect certain features in our daily environment. The gradual onset of daybreak, the brightness of the morning, the gradual waning of light at dusk are vital cues to our inner clock. When they are mistimed or missing, we can suffer from disturbed sleep, depressed mood, and general malaise. Another example: Nutritionists tell us our metabolism expects a diet low in salt and sugar. Both of these were scarce luxuries until two or three centuries ago. Today we have cheap, easy access to them. This has allowed an over-indulgence that is reflected in the dangerous rise of high blood pressure, obesity, and diabetes.

Our world was drastically changed, and continues to be changed, by industrialization and the unchecked growth of cities that followed. Industrialization obeys its own inner laws of growth. But beyond that, one reason it spreads is that it offers obvious advantages. For most people, famine and plague are no longer constant dangers. The average life span has lengthened to an extent unimagined in an earlier day. Access to education has spread beyond a tiny group of elites, and ordinary people have dared to hope that they can make their lives better in the future.

At the same time, the costs of industrialization reach far beyond teeming slums and dark, satanic mills. Today, most people in the Western world are cut off from easy access to the natural environment. They live and work indoors, in artificially controlled climates and under artificial lights. Their daily rhythm is determined more by the needs of the economy than the cycles of the sun.

In theory, those who live in suburbs or in the countryside can still have more natural lives, but in practice this rarely happens. Non-urban areas are increasingly contaminated by water and air pollution. A car-centered culture replaces fields and forests with freeways and malls. Even weather trends are being transformed as a direct result of industrialization. And this pattern is spreading with explosive speed to the rest of the world as well. The consequences of this widespread deprivation are just beginning to be recognized. Fortunately, so are some ways it can be corrected and overcome.

A Healing Strategy

Even before the Industrial Revolution, people were dreaming of "getting back to nature," of returning to some half-mythical Golden Age. Like many dreams, this one usually filters out the more inconvenient aspects of the goal, such as mosquitoes that carry disease and hailstorms that destroy the next winter's food crop. But even if it could be made as perfect as it can sound, it is still unfortunately a fantasy. Our polluted atmosphere infects Shangri-la as well as Coaltown.

At its core, this dream reflects a crucial insight: We were made for a different sort of life than we are now living. We have put that life beyond our reach by the ways we have altered, and continue to alter, our world. But what if we can identify important features of that life, and then re-create them with the help of technology? For example, all of us need clean water to drink, and we would like it to taste fresh as well. Pure mountain streams are harder and harder to find. But even as we work to protect the few rippling brooks that are left, we can also try to promote effective and affordable filters that make even polluted drinking water both safe and tasty.

Bright light therapy is a good example of a technological solution to an environmental problem. Researchers learned that many problems with mood, energy, and sleep are linked to the inner clock, and that the clock is linked to exposure to daily light/dark patterns. Next they worked to find out how to use light (how much? when? how often?) to properly reset the clock. Then the task became to pass this information on to people who need it. Our book is part of that effort.

When we talk to people about bright light therapy, their response is usually, "Yes, of course!" Our intuition tells us that we need sunlight, that it is good for us. Look at people's faces when they step outside on the first bright spring day! Yet our intuition can only take us so far. What if the idea came to us that hearing more songbirds outside our windows in the inner city would calm our anxieties and relieve our depressed mood? We might be right. But before we start agitating for citywide bird feeders, we had best make sure that we have sound scientific evidence to support our hunch.

What are the environmental hazards that make living and working indoors affect our health and well-being? We need to identify and study them, then find practical ways to rectify them. Health science is key to this effort. Itself a product of the Industrial Age, it is still in its infancy. Major scourges of illness and disease still wait to be explored, understood, and cured. Our book looks at only a tiny corner of this effort, but even small building blocks can make a big difference. Our understanding of the health implications of light and dark has evolved dramatically just in the past thirty years, and the correctives we have discovered are almost too simple to believe. This progress shows us that we can make major strides on the health front right now. We do not have to wait around for earth-shaking innovations in microbiology, nanotechnology, or genetic engineering.

Something in the Air

What is so special about spending a day at the beach? What draws tourists and honeymooners to Niagara Falls? Why does the air smell so fresh right after a thunderstorm? We can imagine lots of likely answers to these questions. If you are at the beach, you are not at work (unless you're a lifeguard or cotton-candy hawker). Niagara Falls is a stupefying spectacle, a wonder of nature, and pretty scary if you think about being swept into it. Rain makes the grass smell sweeter and washes the grunge off the concrete. And any of these experiences is enhanced if you are with someone you like. There is something else they have in common, too, though. Each is a situation in which you are likely to find higher concentrations of *negative air ions.*

Negative air ions appear when a molecule of oxygen in the air takes on an extra electron, giving it a negative electrical charge. Unless the air is very dry, microscopic droplets of water, which help protect the charge from dissipating, surround the charged molecule. These clusters of molecules are particularly plentiful at the seashore by the pounding surf, next to waterfalls and rapids, after lightning storms, and in the tropical rain forest.

We are also more likely to find negative air ions out in the countryside than in the city. The reason is that so many of the objects around us—radiators, plumbing, house wiring, electronic devices of all sorts—function as *electrical grounds* that attract and neutralize the negative ions. They don't even need to be directly connected to the earth, as long as they are more positive in charge than the ions. The ions come in contact with them and *presto!*—the negative charge is neutralized. The process is just like what happens on a dry winter day when you take off your coat, then reach for a doorknob or light switch. The electrical charge you accumulated instantly discharges and you feel a slight shock.

There is another reason negative air ions are scarce in our homes, schools, offices, and hospitals: climate control. In the winter, heating systems take cool air that is already dry and warm it up, making it still dryer. In the summer, air conditioners take the humidity out of the air as they cool it. And dry air, whether warm or cool, is the enemy of negative air ions. It strips away those protective water droplets and makes the ions more likely to be neutralized.

When concentrations of negative air ions are low, airborne pathogens—smoke, dust, pollen, animal dander, mold spores, bacteria—build up. The problems these trigger range from breathing difficulties and allergic reactions to headache, lethargy, malaise, and even depression. But when a space is suffused with negative charge, whether by a lightning storm, crashing waves, or an electronic ionizer, some of the pollutants in the air become neutralized. They clump together into heavy particles that drop to the floor, clearing the air. Others take on a negative charge and are attracted to grounded surfaces, so they, too, stop circulating.

This is the principle behind ionizing air cleaners. Most of those sold for home use are too weak to have very much effect, but the industrial-strength ionizers used in factory and laboratory clean rooms create an atmosphere that comes close to fresh, unpolluted air, the kind you might go off to find in the mountains or on an unspoiled beach.

But Is It Real?

There is a good chance you have run across mention of negative ions before now. For thirty years or more, articles in popular health magazines have talked up their benefits, though without offering much in the way of scientific support. Negative ions have even become a mini-fad. A quick Internet search turns up an array of dubious commercial products, including "negative-ion-fiber" clothing, energy bracelets, and special volcanic-rock pendants that supposedly harmonize your body's life force by converting heat into bio-energy. Whether anything is correct in those claims, the ions referred to are not air ions. What started as a plausible idea, based on experience and intuition, seemed fated to end up in snake-oil territory.

In view of this, why are we even discussing negative air ions? What do they have to do with chronotherapy, anyway? The answer is a good example of the way chance sometimes furthers the growth of scientific knowledge.

Back in the 1980s, when we and others first wrote about our success using bright light to relieve depression, many in the psychiatric community found our claims laughable. How could it be? It made no sense from what they knew or believed about psychopharmacology and psychotherapy. It had to be a placebo effect—a treatment that improved the patient's condition simply because the patient *expected* it to.

Researchers do have standard ways to check that possibility. One is using what is called a blind placebo control. Some patients get the real treatment, and some get a harmless, ineffective substitute—a placebo. Most important, all the patients are "blind"—they do not know which treatment they got. Likewise, the research clinicians are kept blind to which treatment the patient is receiving, so their natural bias toward the active treatment doesn't subtly influence the patient's response. That's the gold standard "double-blind" procedure. If those who in fact got the real treatment improve and the ones in the placebo group do less well, it is solid evidence that the treatment works.

There is an obvious problem with applying this approach to finding out whether bright light therapy works. The treatment includes sitting in front of a very bright lamp for a length of time every morning. Patients will know perfectly well whether they are getting the treatment or not. They are not blind, after all! So, with no way to do properly controlled clinical trials, there was no way to prove that the procedure helps patients. Or was there? This is where negative air ions came into the story.

Dr. Larry Chait, a drug researcher, noticed the way negative ions were being touted as a natural mood lifter. But he knew that people have no real way to know whether their air is rich or poor in ions. Since people would believe from what they had read that negative air ions would improve their mood, why not use an ionizer as a placebo control for light—but secretly disable the electronics? Patients would think they were getting active treatment but would be responding only to a placebo.

Experimental Proof

Taking Chait's cue, Rush University researcher Charmane Eastman randomly assigned people with seasonal depression to get either light therapy or sessions with a deactivated ionizer. They informed the patients that some ionizers would be turned on and others turned off, and that it would be impossible to tell the difference. Prior to the treatment, patients expected that the ionizers (assuming they were turned on) would produce just as good results as the light boxes. In fact, however, the light therapy proved superior to the deactivated ionizers.

Soon afterward, we followed up on this approach at Columbia. We divided patients with winter depression into three groups. Group 1 received bright light therapy, Group 2 received high-density negative air ions, and Group 3 received low-density negative air ions. We expected that Group 1 would show more improvement in their symptoms than either Group 2 or Group 3. We were wrong. Those in Group 2 got just as much relief as those in Group 1, while hardly any patients in Group 3 responded to treatment.

The improvement was not just in mood, but in energy, work productivity, sleep quality, and even libido. One patient, George K., twenty-four, joined the study not so much because he was depressed each winter, but because he became completely impotent from November through April. Not only did his depression lift after treatment with high-density air ions, but his sexual performance returned. We should stress that this is a single unusual case. We are not suggesting that others might get a similar result. However, we did notice that on patient questionnaires before and after treatment, ratings of "interest in sex" went up, as it often does when a depression remits. (On the other hand, it is well known that some antidepressant drugs suppress libido even when they are otherwise effective.)

In a follow-up study, we tested patients who had nonseasonal, chronic depression to find out whether the light and high-density ion effects would also work for them. It did. To us, the conclusion from the evidence so far is convincing: *Treatment with high-density negative air ions actively combats depression.*

Andrea's Story

Andrea M. is a twenty-three-year-old graduate student at Hollins University in Roanoke, Virginia. A couple of years ago, while she was still an undergraduate at Hollins, she took part in a clinical trial testing the effects of negative air ions on SAD. The trial was carried out by Professor Randall Flory as part of a long-term research program on treatments for SAD.

"There was a survey for signs of SAD offered to everyone on campus," Andrea explains. "And I decided to take it. I'm originally from up North, and I always had trouble in the winter. It was a problem getting up in the morning—I had no energy. I was not motivated to do anything. It was a struggle to study, do homework.

"I knew I hated winter back in high school or middle school," she continues. "I moved down to Virginia because I was tired of it. It was hard to exercise. I'm very active, but I had no energy to exercise. The winter was dark and dreary. I was very apathetic. Everything was 'Oh, yeah, well . . .' I kept putting things off, you know, like stuff around the house."

Andrea was one of those Professor Flory asked to take part in the study after reviewing her depression score and making sure she was a willing participant. "Every morning, you would come in for either a half hour or hour session with the ionizer. We were not told which group we were in—experimental or control. You had to be within two feet of the ionizer and wear a wristband and sit there for an hour. You could read, do anything you want, but you couldn't use anything that plugs in.

"I had to be there at six or seven in the morning," she recalls. "So in the month of January it was torture getting out of bed. While you're sitting in there, I'd get really sleepy. But after a week or so, I was getting up before the alarm. That was just crazy. After the session, I had this energy when I went back. I didn't feel like sleeping in class. I was taking an environmental studies class, and we went out hiking in January. And I knew it was going to be a challenge, but I fell in love with hiking in January!"

Once the study ended, Andrea stopped using a negative air ion generator, until last winter. "I don't know why last winter was really so tough," she says. "I had a lot of energy trouble. I was also miserable at my job. It was probably a combination of the winter and the stress. And I had no ionizer! So I broke down and got one. I'm a poor graduate student. I wouldn't have spent a hundred and fifty to two hundred dollars on it if I didn't think it worked. I felt I had really seen the results. I use it in the winter all the time, but not as much in the summer. It depends on if I feel I need it."

Asked if she expects to go on using the ion generator for the rest of her life, Andrea replies, "I don't see how I wouldn't use it as I go on—it works! SAD drives me crazy. Seven or eight months a year, I'm a go-getter. I'm doing what I want, I'm active, and then I hit a brick wall. I have the feeling that I know I should do this or that, but I just want to take a nap. I lack motivation to do a lot of daily things: clean, and study. I know I have to get out, get fresh air, get exercise. It holds me back.

"I always thought I had SAD," she concludes. "I wanted to be in the study because I thought it was time to find out if I really had it. When I told my mom and dad, they were like *'What?!'* It seemed completely crazy and silly to them. But you have to be open-minded—because now they see that it works."

How Does It Work?

When we tell people about negative air ion therapy, one of their first questions is "How does it work?" At this point our best answer is, No one really knows. We do realize how unsatisfying that is. We, too, find it frustrating. Clinically, it may not matter so much, as long as the treatment has been shown to be safe and effective. As we mentioned earlier, around 400 BC, the Greek physician Hippocrates described using the active ingredient in aspirin to treat pain and fever, but it was not until the 1960s (AD!) that scientists finally discovered how it works.

We hope it will not take twenty-three centuries to understand the effects of negative air ions. There are a few ideas around already. Early lab studies pointed to an effect on the brain chemical serotonin, but results were inconsistent. Another possibility is that negative ions spread over the skin surface, neutralizing any positive charge on the body (assuming that such positive charges cause depression). Or maybe it is the inhalation of the ionized air that matters. The ions might activate the vomeronasal organ—a mysterious structure in the nostrils that is thought to detect pheromones—and send a neural signal directly to the brain.

Our preferred hypothesis for now is that when ionized air reaches the lungs, it improves the uptake of oxygen into the blood. One study that pumped negative air ions into the lungs of postoperative patients in intensive care found that levels of stress-related lactic acid in the blood went down. As we write, a study is under way to follow the level of blood oxygenation in SAD patients during high-density negative air ion treatment. If this proves out, we may be able to point to increased blood oxygenation as having antidepressant action.

Some Practicalities

For those who are considering whether negative air ion treatment may benefit them, an important point to consider is the ionizer itself. Quite simply, most of the devices being sold are too puny to bother with. Even if they do produce negative air ions (and we've seen some that do not), the concentration they produce is so low that it is unlikely to have any clinical effect. The industrial-strength ionizers we have used in our research raise the concentration of negative ions in the air by a thousand times or more (see *Resources for Follow-up*, p. 301).

Assuming you have a powerful ionizer, what then? One approach we have used successfully involves creating a *closed loop.* It is important that the user be well grounded, so that the negative ions are more attracted to the user than to the TV, the radiator, or the wall outlets. To do this, we use a conductive bracelet attached to a wire that leads to the ionizer. The ionized air flows to the body and any excess charge passes back through the wire to the ionizer's ground. We have found that sessions as short as thirty minutes in the morning are effective, and the person's mood generally improves within a week or two of daily treatment.

Another approach involves ionizing the whole room. The windows and doors are shut and the ionizer is turned on. After about an hour, the negative air concentration builds up to *room stasis*, the point at which negative ions are being replenished just as fast as grounded objects are depleting them. This method has been used successfully for overnight exposure in the bedroom while the user sleeps. Nighttime exposure can be cut to as little as ninety minutes and still produce the antidepressant response by using a special conductive bedsheet that is connected by a wire to the ground of the ionizer.

Some Cautions

There are few cautions about the use of negative air ionization. So far, we have seen no clinically significant side effects. There is, however, one case report: A psychiatrist self-treated and woke up after a night's exposure feeling so energized that his mood verged on hypomania (an exaggerated euphoria). We should point out that *all* antidepressants, whether drugs or lights or air ions, have a small risk of inducing hypomania. Reducing the dose can usually control this. In any case, caution is always a good idea.

Some ionizer designs also produce significant levels of ozone, a noxious substance that is odorous and can cause stinging in the throat. Any ionizer used for therapy should be designed to minimize ozone production. We have seen deceptive commercial claims for ionizers that are in fact quite noxious and exceed the limits set by agencies like the U. S. Food and Drug Administration. With the well-designed devices we use in the lab, this problem has never occurred during thirty- to ninety-minute sessions. However, if ozone is detected during overnight or twenty-four-hour use, the windows and doors should be opened slightly to increase air circulation. Again, caution is indicated.

Patients who want to increase their chances of feeling better often ask if they can use light therapy together with air ion therapy. So far, no one has clinically tested this sort of combination treatment. There are commercial devices that combine the two technologies, but we have doubts about their value. If an ionizer is encased in a light box, or placed close to a light box, most of the ion flow will be attracted to the light box ground, not the user. In that case, we would not expect the negative ions to have any effect. If a user wants to try both methods, they should be done at separate times. As far as we know, the timing of negative air ionization is not critical in the same way as the timing of bright light therapy.

Here we have an environmental fix—separate from chronotherapy—that both creates healthier air for breathing and lifts the burdened soul.

PART 4

Stages of Life

11.

The Promise of Pregnancy

Having a baby is one of the most natural things in the world. Every single one of your ancestors went through parenthood with at least enough success that you are here to tell the story. We are as well prepared for this life phase as our biological and genetic heritage can manage, and that is very well indeed. At the same time, however, pregnancy brings with it both complexities and vulnerabilities. For the developing fetus, these complexities include the formation of the brain, eyes, and nervous system. For the mother, they may include mood and sleep problems, both during and after pregnancy. Chronobiology gives us new ways of understanding all of these, and chronotherapy offers new ways of dealing with them.

The Developing Inner Clock

In our grandparents' day, watchmaking was a very complex and delicate affair. Hundreds of tiny parts had to be fitted together precisely. Only when all were properly in place did the watch start ticking and keeping time. The story is similar with the circadian timing system of the developing fetus. Each of the important elements—the eyes, the pineal gland, the suprachiasmatic nuclei (SCN) in the brain—develops at its own pace. All need to reach the point of being able to function and communicate together before the system can do its job of controlling daily cycles such as sleeping and waking.

The eyes come first. They begin, just four weeks after conception, as two thickened spots at the base of the brain. Gradually they grow out on stalks, which will become the optic nerves, and move toward the front of the head. Meanwhile, the various parts of the eye differentiate, and the eyelids seal closed to protect the light-sensitive structures. After twenty weeks, all the eye components have developed. By thirty weeks, the baby's eyes are wide open much of the time, and a week later, the pupils widen and narrow, showing that the retinas are able to detect light entering the eyes.

The pineal gland starts to develop next. Unlike the eye, it does not emerge from brain tissue. It is also one of the few structures in the brain that is not protected by what is called the blood-brain barrier. Buried deep in the center of the brain, it is only about as big as a grain of rice at full growth, but the melatonin it produces is crucial to regulating the timing and synchronization of the body's rhythms.

The SCN—the inner clock of the brain—has yet to appear. It has to differentiate itself from the larger structure of the hypothalamus, and that does not finish happening until late in the second trimester. This is much later than the development of either the eyes or the pineal gland. The SCN starts affecting body temperature, release of the activating hormone cortisol, and other daily cycles while the baby is still in the womb. However, this circadian pattern is still not linked to the outside world except by way of signals from the mother.

In Touch with the Fetus

The pregnant mother and the developing fetus are biologically linked by the placenta and the umbilical cord. These structures pass food and oxygen from the mother to the baby, and pass waste products and carbon dioxide from the baby to the mother. This process is well known, both to the medical community and to parents who have taken birth classes or read up about pregnancy. What is much less widely recognized is the way these structures allow the mother's circadian clock to affect the baby.

The placenta forms the boundary between the mother's blood system and the baby's, and like boundaries between countries, it has a sort of customs station that keeps some items from crossing the border. This *placental barrier* works well in blocking many harmful substances, such as bacteria, from reaching the fetus (though others, such as alcohol, some drugs, and some viruses, do get through). At the same time, the hormones in the mother's bloodstream get a green light. This is the most important means by which the mother's circadian rhythm is passed along to the fetus.

In simple terms, the daily sleep/wake cycle closely follows the alternating rise and fall of two hormones: melatonin from the pineal gland, and cortisol from the adrenal gland. During the sleep period, melatonin levels are high and cortisol levels are low, and during the wake period it is just the opposite. Both these hormones cross the placental barrier and reach the fetus, letting it "know" when it is wake time and sleep time for the mother. Cortisol levels also shoot up when the mother is under stress, which is a good reason for trying to stay as relaxed as possible.

The mother's daily rhythm is communicated to the fetus in other ways as well. Her meal pattern determines the timing of the flow of nutrients through the placenta. As the mother's body temperature goes through its regular daily cycle, it creates variations in the environment of the fetus. So does the contrast between the mother's active movements during the day and the relative lack of motion when she is asleep. Probably in response to these variations, the fetus is much less active in the early morning than late in the evening.

This maternal signaling is essential for the development of the baby's own circadian rhythms during pregnancy. It also sets the stage for the baby's life after birth. The more regular the mother maintains her own rhythm during pregnancy, the more the baby benefits. It will be half a year before the baby's circadian system matures enough to independently sync with the outside world, but it starts functioning on its own, with an approximate twenty-four-hour cycle, within one week after birth.

One indication that these time cues from the mother are so important is the case of babies born prematurely. They stop receiving the maternal cues at birth, of course. Meanwhile other fetuses of the same conceptual age go on getting the cues until they, too, are born. Apparently as a result of this, premature infants take longer to establish a regular sleep pattern than infants born at full term.

Since we mention premature babies, we should also raise a related point. Researchers have recently studied hospital nurseries and infant intensive care units where premature babies must often be cared for. They found that often these facilities are brightly lit around the clock. What this lack of a light/dark cycle does to the newborn's inner clock is not yet clear. Even so, experts say that placing these very vulnerable babies in a more deliberately circadian environment will benefit their development.

Some Steps to Take

What can you do while you are pregnant to give your baby a healthy circadian start in life? Here are a few points to consider:

- Eat breakfast soon after waking up, and eat your other meals on a consistent schedule.
- Get enough sleep, and get it during the same periods daily, including on weekends. This will help keep your inner clock regular.
- Avoid strenuous activity before bedtime that might artificially raise your body temperature.
- If you have trouble waking up in the morning, use morning light therapy to shift your inner clock earlier. This can also be a mood-lifter!
- Cut down on or eliminate non-essential medications, which can pass the placental barrier with unknown consequences for your baby, including possible effects on the baby's developing circadian system.
- If you have other children, ask your partner to deal with any nighttime calls if they're having sleep problems of their own.

Bear in mind, we mean these as *precautionary* measures. Babies are resilient creatures—they can reach a healthy full-term birth and enjoy a healthy infancy even when conditions are far from ideal. The steps we are suggesting are meant, essentially, to give them that much more of a chance of healthy development.

Mood Problems During Pregnancy

Many people like to think that being pregnant is a guaranteed source of joy and contentment. In the media, a pregnant woman is often shown holding baby clothes and gazing out the window with a dreamy expression on her face. Of course many women who are pregnant do feel exactly that way, at least some of the time. But when they don't, when they sometimes find themselves plunged into feelings of sadness, worry, or guilt, they start to think there must be something wrong with them. Aren't they supposed to be happy?

In fact, mood problems during pregnancy are widespread. As many as one in eight women have to deal with clinical depression while they are pregnant. Many more experience burdensome periods of feeling sad, blue, helpless, or deeply tired, short of clinical depression. These are very real problems that can have serious consequences for both the woman and her baby. Yet many of them go undiagnosed and untreated. Women often assume that they are merely going through a typical and temporary moodiness, and their doctors often agree.

This is a distressing situation. Mood problems during pregnancy can be safely treated and effectively relieved by chronotherapy as well as other means. Ignoring or dismissing them is cruel to the woman and damaging to her baby.

What to Watch For

During pregnancy, most women feel more tired than usual at some point. Most have problems sleeping from time to time. And yes, most have moments of feeling worried or blue or even helpless. So how do you know if your mood problems are serious enough to turn for help?

Here are some of the most common ways depression shows itself. If you experience any of these for two weeks or more, you should ask your health care provider about how to get help.

- You lose interest in activities you used to enjoy.
- You pull away from family members and friends.
- You feel sad or blue most of the time.
- You have no energy or motivation.

- You cry a lot.
- You sleep a lot more or less than usual.
- You eat a lot more or less than usual.
- You can't concentrate or make decisions.
- You are unusually irritable, argumentative, or anxious.
- You feel worthless or guilty.

For a more detailed look at your emotional state and what it may mean, take our online confidential self-assessment of depression severity (see *Resources for Follow-up*, p. 301). This will give you specific personalized feedback and advise whether it is important to discuss the problem with your OB/GYN.

When someone is experiencing several of the symptoms in the list at the same time, it is considered a *major depressive episode*. The pattern of symptoms may be different from one person to another. If the depression becomes intense, a person may also have intrusive thoughts that life is not worth living, thoughts about or visions of dying, or imagining or planning self-harm. At times like these, it is hard to believe that this state is temporary and that you will feel better. Light therapy provides one of the safest ways to spring out of this situation, but in this case it should be used only with your doctor's monitoring.

If you develop feelings like this that are vague but disturbing, it is important to tell someone close about it or to speak confidentially with volunteer experts who can guide and assist you (check the international directory at befrienders.org). If these feelings become so strong that you are uncontrollably focusing on your own demise, you or someone you trust needs to take immediate action. Go to the nearest emergency room or dial your local emergency number (911 in the U.S.). Help is there to calm these intense thoughts, but it depends on you to seek it.

Mood Problems Affect the Fetus, Too

Many women when they become pregnant experience a higher degree of physical and emotional stress. The changes in their bodies and their hormonal systems thrust them into new, often upsetting, territory. That can happen whether or not they also become depressed. But depression multiplies the ordinary strains of pregnancy. This has consequences that harm both the woman and the developing fetus.

Women who are depressed during pregnancy have trouble summoning up the energy and desire to give themselves the attention they need. They are likely to neglect their diet and get inadequate nutrition, at a time when their nutritional needs are greater and more critical. As a result, they do not gain as much weight as a healthy pregnancy demands. They are also more likely to use harmful substances such as tobacco, alcohol, and illicit drugs while pregnant. The sense of hopelessness that is so often part of their emotional state makes it harder for them to show up for appointments with the doctor or prenatal clinic and harder to follow important medical instructions.

As for the baby, depression during pregnancy has been linked to lower birth weight and premature birth, both of which pose dangers to the baby's health. Depressed mothers are more likely to need surgical help during delivery, and their babies are more likely to need to be in the neonatal intensive care ward. And recent research has shown that newborns with depressed mothers have reduced muscle tone and higher levels of stress hormones in their bloodstream.

Being depressed during pregnancy is also the strongest predictor for being depressed after childbirth. This carries its own set of risks for both mother and baby. Understanding these risks, the ways they may be linked to problems with the mother's inner clock and the ways chronotherapy can help, is an important preventive step for pregnant women as well as their loved ones and doctors.

Marcia's Story

Marcia C., thirty-one, lives in a suburb of New Haven with her husband and six-month-old son, Gianno. She is a yoga instructor and switched to working part-time after Gianno was born.

Marcia has had a delayed sleep phase problem since she was eleven. "Basically my pattern was to sleep from three AM to eleven AM," she explains. She adds that her husband is an early riser. "He's asleep by ten o'clock. There were hardly any hours of the day that we could see each other, because when I was working hard in the evenings, he would go to bed before I got home and I'd be up till three AM. I never used to see him in the morning, except for when we were traveling and then I was grumpy [because of lack of sleep]."

Over the years Marcia had tried both therapy and medication to deal with her sleep problem, but without much success. She decided to try again when she was going to have a baby. "I read an article about chronotherapy maybe a month before I was pregnant," she recalls. "I knew I would be at such a huge disadvantage if I didn't change my sleep pattern. I thought I'd better do something immediately because having to wake up for a baby, I would end up with very little sleep."

When Marcia came to the clinic, we explained that she should do the light therapy at the natural end of her sleep cycle, which at that point was in mid-afternoon. "When I did the light the way Dr. Terman suggested, I would wake earlier and earlier in small increments. I couldn't believe how *easily* it worked. I was cautiously optimistic. But I didn't want to get too hopeful. After a month, or a month and a half, I was going to sleep at midnight and getting up at eight AM. Of course I was getting up to pee throughout the night, but I had no trouble getting back to sleep."

Marcia's treatment officially finished four or five months before Gianno was born. "But I kept using the lights, because it really worked. When Gianno came, I couldn't maintain it for quite a stretch, but when he was on more of a schedule, I talked to Dr. T. He gave me some general guidelines so I could go to sleep by eleven and get up at seven. When I was first using the light again, Gianno wasn't even close to napping. I had to figure out how to sit without his getting the light. I put a sun hat on him, and I put him in the Bjorn, and I sat on a yoga ball.

"During the first few months after Gianno was born, I was a *mess*," she adds. "I didn't know how much was hormonal. He was fussing, it was overwhelming, and it was also winter, so I was not leaving the house a lot. I think the lights work in general—with sleep issues and psychological issues. My belief is, I get depressed from sleep deprivation, especially when it's combined with stress."

Marcia can easily see herself continuing with light therapy for the rest of her life. "It may be burdensome, but being able to maintain healthy sleep far outweighs the burden. It's like night and day. I never imagined it could have happened. I spent so many years *alone* in the evening, *alone* at night, no one else around. Joe and I had never gone to bed at the same time, and I didn't realize he twitched! He wound up going to see someone for his twitching. For all these years—four years—I didn't know how he slept. I feel more connected to my husband and to my life now."

Treating Depression During Pregnancy

The two most common approaches to treating depression in pregnant women (as with people in general) are psychotherapy and antidepressant medication. Both present difficulties. Psychotherapy is not easily accessible to most pregnant women who are depressed. This is especially so for those who are poor or who live outside major urban centers. Even when it is available, practical and emotional problems add up to make them more likely to drop out of treatment.

Antidepressants pose a different problem. They are certainly easily available, as close as the nearest pharmacy. But many pregnant women and their care providers hesitate to use them. All antidepressant medications cross the placental barrier and reach the baby. Is that harmful? This is a highly controversial question. Antidepressants are said to be safe in general, but there are few well-controlled studies of their effects during pregnancy. Many of them work by affecting the neurochemical serotonin, which plays an important role in the development of the baby's brain. One recent large-scale study found that babies exposed to these drugs were likely to have reduced head growth. They were also more likely to be born prematurely.

On the other hand, women whose depression is *not* treated are also more likely to have babies with smaller heads (as well as smaller bodies) and to give birth prematurely. If a pregnant woman had to choose between taking antidepressant drugs for her depression and not treating her depression at all, she would need to study the balance of risks and benefits very carefully. But is that her only choice? We don't think so.

Light therapy has proven itself to be effective and safe in treating both seasonal and nonseasonal depression. For years, however, researchers were discouraged from studying its usefulness with pregnant women. Funding agencies and ethics committees were concerned that it might not be safe for the fetus. (Women have been spending their entire pregnancy out in the sunshine for countless generations, but this was not considered evidence.)

To meet these concerns, we did a small pilot study with depressed pregnant women. Five weeks of light therapy made a marked improvement in their depression. Their babies all gained weight at a normal rate, were born normally, and showed no ill effects from their mothers' light treatment. This evidence allowed us to collaborate with Swiss researchers on a major five-year clinical trial. The results were published in 2011 in the *Journal of Clinical Psychiatry.* By the end of five weeks of treatment, 83 percent of the women who received bright light therapy showed major improvement, as compared to 46 percent of those who used a placebo light box.

Of course there is a need for further research. There is almost *always* a need for further research! But in our opinion, enough is already known about the effectiveness and safety of light therapy to make it a prominent choice for pregnant women who are suffering from depression or mood problems. It is simple, inexpensive, and biologically based. The side effects are infrequent and minimal, there is no known risk to the unborn baby, and neither mother nor baby is exposed to powerful, potentially harmful drugs.

After the Baby Is Born

"Both mother and baby are doing well." These are the words everyone wants to hear at the conclusion of a pregnancy. What they hope they will *not* hear is some variation on "Well, the baby's fine, but Mom's been looking a little down in the mouth." Most people, especially those who are pregnant or who are close to someone who is pregnant, have heard stories about the "baby blues." They know about the unsettled feelings and mood swings that so many women experience during the first week or two after giving birth. As a rule, these quickly pass. But not for everyone. For some, the unhappy, negative feelings they are going through are severe enough to qualify as clinical depression. Because it happens after childbirth (in Latin, *post partum*), this is known as postpartum depression.

In most ways, postpartum depression is very similar to other depressions, whether they occur during pregnancy or at other times in life. The new mother may feel anxious and deeply sad, become agitated or irritable, find it hard to do routine tasks and far too easy to burst into tears. There are differences, though. Depressed new mothers are often overwhelmed by feelings of guilt about *being* depressed. How can they take care of their baby if they can't even take care of themselves? They may even find themselves tormented by thoughts (rarely acted on) of harming the baby.

Effects on the Baby

Being depressed is a miserable experience for the mother, but it is even worse for the baby. Ordinarily, during the first weeks and months, babies and caregivers—mothers, fathers, and others—engage in an intricate reciprocal exchange of gazes, facial expressions, and vocalizations. The baby meets your eyes, you coo or make a funny face, the baby gurgles and smiles, and the game continues. This is a vital element in creating an emotional attachment on both sides. It is just as vital in helping the baby start to develop an understanding of the world it has been born into.

These crucial interactions happen less, and happen differently, when the mother is depressed. Depressed mothers smile less at their babies, touch them less, talk to them less, sing to them less, look at them less, and play fewer games of peekaboo with them. When they do touch the baby, they do it with less affection, and when they do talk to the baby, they are less responsive to the baby's reactions.

The effects of all this on the baby are far-ranging. By the age of five months, babies of depressed mothers are less likely to notice the difference between a neutral and a smiling face. They pay less attention and learn less easily when someone speaks to them. When depressed mothers do try to read to their babies, the babies are more likely to push the book away or try to shut it. All these behaviors help confirm the conclusion that babies with depressed mothers form less secure attachments. And evidence continues to mount that babies who develop attachment problems can go on being affected by these all through childhood and adolescence and into adulthood.

When and Why It Happens

As a rule, if someone is going to develop postpartum depression, it happens early, either very soon after childbirth or within the first three months, though occasionally it can appear as much as a year later. One of the triggers seems to be the shock to the system when the level of hormones that peaked at the end of pregnancy—estradiol, prolactin, and cortisol—suddenly drops. Hormones produced by the thyroid gland may also drop sharply, contributing to feelings of tiredness and moodiness. In contrast, inflammatory cytokines, which are helpful in the run-up to delivery, may stay at too high a level. Even if the first weeks pass without problems, other factors may come into play, including the psychological stress of having the primary responsibility for the baby's well-being.

Then there is sleep. If a woman experiences poor sleep toward the end of pregnancy, even if she is *not* depressed, it is more likely that she will become depressed a month after birth. Poor sleep during the period after delivery indicates a higher risk for either a first-time episode of depression or a relapse for those who have known depression in the past. Women who find themselves napping during the day are also at greater risk. Napping, of course, is often a consequence of poor sleep the night before, and once the urge to nap hits, it can be virtually impossible to suppress.

Clinicians have noticed that mothers have better, more regular sleep once their babies develop a more predictable sleep/wake/feeding schedule of their own and once the feedings become spaced further apart. And for some, at least, once their sleep improves, their depression lifts as well.

All this points to the need for careful sleep management both before and after the baby is born. A preliminary study in Seattle, Washington, started light therapy with a group of new mothers who had become depressed and had not been treated during pregnancy. The improvement was impressive, but so was the improvement using a dim red light thought to be a placebo. Thus, the study cannot be considered definitive and successful.

We have used light therapy with women suffering from depression during pregnancy. In response to the treatment, not only did their mood improve, so did their sleep. That alone may have reduced their chances of postpartum depression. Unfortunately we could not extend our study into the period after childbirth, but several women decided on their own to continue using the light treatment after returning home from the hospital. They said it kept them going.

Until there are clinical trials that provide scientific evidence for the effectiveness of light therapy for postpartum depression, we cannot be sure if our patients avoided postpartum depression *because* they continued the light therapy. But in the meantime, we think it makes very good sense for new mothers to follow their example. The procedure is well tested and safe. And yes, you can breast-feed during the light therapy session, if you make sure the baby is nuzzled against your breast and not directly exposed to the bright light signal. At the same time, partners should do all they can to make the mother's nighttime sleep better. If the baby still needs that 3 AM (or 6 AM!) feeding, maybe it should be from a bottle, in the partner's arms, while the mother catches a little more sleep and rebuilds her strength.

12. Strategies for Babies and Children

Advertisers know that a cute baby or child is a way to attract people and spread an aura of positive feeling over a product. Some psychologists say that we are pre-wired to respond positively to certain characteristics of the young, such as a large head, a tiny nose, fluffiness, and a general air of helplessness. Which may be why advertisers also know that if you can't get a baby, a kitten or puppy—or even a tiger cub—will do almost as well. At the same time, though, people are also drawn to stories and films about changelings and weird demon children whose seeming weakness masks uncanny powers and unexplainable malevolence.

Parents sometimes find themselves switching back and forth between these attitudes. Their new baby is the most precious thing in the world to them—until it wakes them for the third time in a single night and will not stop crying. Then they may start to wonder what kind of creature they have taken into their lives. Can chronotherapy offer them any understanding and relief? We hope to show you that it can.

Newborns and Sleep

One thing everyone knows about newborns is that they spend most of their time asleep. On average, that means eight or nine hours of sleep during the daytime and another eight or nine at night. Their stomachs are small compared to the amount of food their rapidly developing bodies need, so hunger generally wakes them every two and a half to four hours. They eat, look around, and interact with whoever is holding them, feeding them, or meeting their gaze. Then they drop off to sleep again.

To new parents it may seem that all that sleeping is simply downtime, when nothing much is going on with the baby. Not so. As much as half the time newborns are asleep, they are in *REM* (rapid eye movement) sleep. In comparison, by the first birthday REM has declined to less than a third of sleep time. REM sleep is light and active. The baby may smile, frown, make sucking motions, and wave arms and legs around. In older children and adults, this is also the stage of sleep in which dreams occur. So far no one has figured out a way to tell if newborns dream during REM sleep, but they certainly act as if they are. And REM sleep is important for the baby's development. It helps learning and memory and promotes the rapidly spreading interconnections among brain neurons.

As parents know all too well, the sleep/wake cycle of newborns is not controlled by the external cycle of light and dark. After they have been asleep for three or four hours, they wake up hungry, whether it is 3 PM or 3 AM. Some premature babies need to eat even more frequently and may have to be awakened for a feeding. Bottle-fed babies may sleep a little longer than breast-fed babies, but that's a result of the digestive system, not the inner clock. During the first five or six weeks, the baby should not sleep longer than five hours between feedings. Parents should be alert for any sudden change in the baby's sleep pattern, which might indicate a problem.

Recent research has found signs of a weak twenty-four-hour sleep/wake pattern in some newborns. This may simply be an echo of exposure to the mother's circadian rhythm while still in the womb. Still, we do know that the biological clock starts being able to affect body temperature and heart rate even before birth. More generally, the baby's twenty-four-hour rest/activity cycle starts to emerge fairly soon after birth and gets stronger across the first six months. This is partly a matter of maturation, but it is also influenced by the way the baby is taken care of.

Since the inner clock is sensitive to light exposure even in very young babies, it makes sense for parents to pay attention and regulate that exposure. This means getting sunshine during the day, of course, and preferably in the morning. It also means keeping light levels low during the night. For those 3 AM feedings, parents need the security of being able to see where they are going and what they are doing, but we strongly suggest putting in amber night-lights instead of turning on room lights (see *Resources for Follow-up*, p. 301). Amber light does not affect the baby's circadian system the way white light, which contains the blue part of the spectrum, would. As a bonus, the amber light does not halt melatonin production in the parent's system. That should make it easier to get back to sleep once the feeding is over.

Through the Night

You do not need a birth announcement to spot parents of a new baby. The bags under their eyes are often announcement enough. After all the stresses of pregnancy and delivery comes the new reality of being on call around the clock. The nightly wake-ups in the middle of the circadian sleep zone seriously tax a parent's system, even when there are two parents to share the burden. When there is only one, the burden is much more than twice as heavy.

No wonder, then, that one of the most frequent questions new parents ask is, "When will my baby start sleeping through the night?"

Unfortunately, the answer is both complicated and highly individual. Before three months of age or so, very few babies are ready to give up that feeding in the wee hours. By the time they are six months old, most are able to do so. That is generally what parents mean by sleeping through the night, and it usually comes as a great relief. However, often the baby is not really sleeping through the night, and this is where problems can arise.

Until their first birthday, most babies sleep in a series of cycles that are each about an hour long. In the early part of the night, they sleep very soundly. Then they move into a lighter sleep that alternates with REM (dreaming) sleep. Several times a night, these REM periods are followed by short intervals of semi-waking. Babies open their eyes, perhaps fuss a little, and go back to sleep, all in five minutes or less.

Often enough, tired parents do not even notice these events. As far as they know, the baby has slept through the night. But some parents may be particularly sensitive to sounds from the baby. Or the baby may be spooked by something, wake up completely, and start crying. Suddenly a perfectly normal sleep pattern comes to seem like a problem. The baby is "not yet sleeping through the night."

Sometimes babies who have been sleeping through the night start waking up again during the night. This often happens after they are around six months old. The reason may be some discomfort, such as teething pain, a cold, or an earache. It may be overstimulation from practicing new skills such as sitting up, crawling, and walking. Many babies this age also develop *separation anxiety.* This is a normal development that takes place when they realize that parents and caregivers are separate from them and can go away—but will they come back?

Letting Them Cry?

If the baby's nighttime sleep pattern becomes a source of concern, many parents start to wonder, "How can I *teach* my baby to sleep through the night?" They will find plenty of "baby trainers" ready to offer guidance. Their aim, they say, is to keep you from spoiling the baby and to keep the baby from manipulating and blackmailing you. As for the techniques they offer, most are variants on the age-old approach of let-them-cry-it-out. Parents are sternly lectured not to let down their guard by responding to their baby's nighttime crying. Supposedly that will only reward the baby's misbehavior and make it more frequent. Next thing you know, you will have to feel guilty that you have created a spoiled brat.

In our view, these approaches are deeply misguided. This is especially so when they are applied to babies under the age of six months. A baby's sleep cycle is at least as much the product of that baby's individual temperament and developing inner clock as it is of the parents' approach to bedtime and sleep. Sensible parents will try to understand the changing pattern of their baby's sleep needs and adjust to them. In any case, a baby's sleep habits should not be seen as a source of bragging rights. Friends may tell you how their baby slept through the night at three months. *So?* Even if their account is not exaggerated, as so many are, *their* baby is not *your* baby. What worked for them may or may not be a good idea for you.

The notion of letting the baby cry it out is wrong-headed for a whole range of reasons. Certainly if babies go on crying long enough, they will eventually fall asleep from exhaustion. But is it reasonable to call that a healthy sleep? On a hormonal level, extended crying leads to higher levels of arousal and stress hormones—particularly cortisol, which directly counteracts the effects of the sleep-related hormone melatonin. It also sets off increased heart rate, blood pressure, and body temperature. Even at normal levels, cortisol works to get the body ready for waking activity, not for sleep. At the elevated levels produced by distraught crying, it practically guarantees that normal sleep simply will not happen.

We should also remember that during the first year or so, crying is one of the few ways babies can communicate. It is their way of letting the world know there is something they need—food, a dry diaper, some ointment on that rash, a reassuring hug. . . . If the world responds by ignoring them and turning away, the message is clear: Whatever it is you need, it does not matter to us. They may eventually give up crying at bedtime or during night wakings, but this is more likely from despair than from some newfound maturity. The state they have been thrust into is like what psychologists call *learned helplessness*, and that in turn has been linked to depression. And yes, depression can strike infants as well as children, adolescents, and adults. Understanding its sources in social as well as chronobiological factors is a crucial first step toward effectively dealing with it.

Older Babies, Toddlers, and Preschoolers

A baby's sleep/wake cycle gets much more organized over the first few months of life. During the night, sleep consolidates into fewer and longer periods, while daytime sleep gets shorter and turns into well-defined naps separated by longer waking hours. By the age of six months, babies typically sleep for two five-hour stretches at night, with a single feeding in the middle. They also take morning and afternoon naps of one or two hours, for a total of around fourteen hours a day. By the first birthday, the total is still around fourteen hours, but naps are shorter, nighttime sleep is longer, and the midnight feeding fades away.

We should stress that all this is based on that mythical creature, the average baby. Overall, a baby's sleep needs and the shape of the sleep/wake cycle are largely determined by age and stage of development. However, individual temperament and a family's usual schedule and routines can create wide differences. Parents who become concerned that their baby is sleeping too much or too little should talk it over with the baby's physician.

Across the toddler and preschool years, the amount of sleep the average child needs slowly declines, from the one-year-old's fourteen hours, to thirteen hours at age two, twelve hours at age three or four, and eleven hours at age five. Napping gets shorter, then disappears. The morning nap goes first, followed by the afternoon nap. Between three and four, the majority of kids give up napping, but many go on taking naps until they are five or even older.

Once children give up naps, their nighttime sleep usually gets longer to make up for the lost daytime sleep. The shape of their nighttime sleep changes as well. The first third of the night is mostly deep non-REM sleep. REM sleep comes mainly in the second half of the night, and there is less and less of it. Recall that newborns spend half their sleep hours in REM. By age five, REM sleep has declined to about 20 percent, which is not much different from the proportion found in adult sleep.

School-Age Children

Sleeping patterns continue to change during the school-age years, from five to ten, but not nearly as dramatically as in the preschool years. At five, the average child needs around eleven hours of sleep, and by age ten this has declined to slightly over ten hours. Unfortunately, quite a few school-age children get less sleep than they need. The many reasons for this include caffeine, evening TV watching, computers in the bedroom, and a general sense that life is much too exciting to shut it out by going to sleep.

Bedtime anxieties often play a role as well. By age five or six, kids are aware enough to start taking in the scary stories they hear on the evening news. However, they do not yet have the perspective to understand that a tsunami in East Asia or a kidnapping in a distant part of the country is not a threat to them personally.

This is the age at which a child's chronotype begins to show itself more clearly. Just as with teens and adults, some children are larks, some are owls, and most are hummingbirds, somewhere in between. Whether this is mostly biological, or a question of earlier experiences, or some combination of the two, it is a fact that parents will need to deal with. Some kids consistently wake up early, bright and bushy-tailed and ready to go. Others get off to a slower start, not really reaching top gear until sometime in the afternoon. It is a wise step for parents to notice these propensities and to try to organize the child's day in line with them.

A child's chronotype can become an issue at bedtime, too. Suppose you find out that most of your second-grader's classmates go to bed at 8 PM and get up at 7 AM. That works out to eleven hours, which sounds about right, so you decide to adopt that routine. If your kid happens to be a hummingbird, that may work out fine. But a lark's inner clock may signal for sleep sooner than 8 PM, and give the wake-up signal at 5:30 or 6 AM. That means you have an already over-sleepy kid at bedtime who is sleep-deprived in the morning. Owls are even worse off. Bedtime is 8 PM, but circadian sleep time is later, say 9 or 9:30 PM. It's not that these children are resisting going to bed; they simply are not biologically ready to sleep. Result: an over-alert kid at bedtime who is groggy in the morning.

It would be nice if we could offer a guaranteed solution to problems like these. Unfortunately, schools, school buses, and parents' obligations have schedules that pay no heed to individual chronotypes. If kids have to be up by 7 AM to get to school, that's that. It's easy for larks, okay for hummingbirds, and hard on owls. Parents can try moving bedtime earlier for their larkish children, but they can't insert extra hours into the nighttime for their owls. All we can suggest, realistically, is that they try to find a compromise between their child's natural sleep time and a bedtime that will leave room for an adequate amount of healthy sleep. We outline some approaches to doing this later in the chapter.

Some Sleep Problems of Children

Most children experience sleep problems at some point between infancy and adolescence. By "sleep problem" we mean a pattern of sleep that keeps the child from getting enough healthy sleep or that seriously disturbs the sleep of other family members. Some of these, such as night terrors, usually happen only rarely and are limited to a particular stage of development. Others, such as bedtime resistance, practically become routine in some families.

Sleep scientists divide sleep problems into two groups: *dyssomnias* and *parasomnias*. Dyssomnias are problems that interfere with going to sleep or getting enough sleep. An example would be those plaintive, repeated bedtime cries of "I'm thirsty," alternating with "I need to peepee." We discuss some approaches to dealing with them in the next section.

Parasomnias, in contrast, are neurologically based events that break into ongoing sleep. Three of the parasomnias—confusional arousal, sleepwalking, and night terror—have a great deal in common. All tend to happen during the first third of the night, during deep non-REM sleep. As a rule, the child stays asleep while they are going on and does not remember them the next morning. And, scary as they may be for parents, they are basically harmless and likely to disappear as the child gets older.

Confusional arousal is the least startling, so much so that parents may not even notice it. Children may sit up, mumble or call out, look around in confusion, and thrash about. Then after a few minutes they fall back into a deep sleep. Children who are *sleepwalking*, on the other hand, often get out of bed, wander around the house, eyes open, and even try to go outside. Almost half of school-age children sleepwalk at least once or twice, and some do it regularly.

For the parents of frequent sleepwalkers, the first concern should be keeping the child safe. That means locking outer doors, blocking off hazardous staircases, putting dangerous objects out of reach, maybe even fastening a cowbell to the child's bedroom door. By the way, forget that old story about the dangers of waking a sleepwalker. It is *not* dangerous, but it is fairly difficult as well as pointless. A better move is to gently guide the child back to bed.

Any parent who has been through a child's *night terror* retains a vivid memory of it. The child suddenly wakes up—or rather, *seems* to wake up—screaming. Attempts to comfort or calm the child only seem to set off a worse panic. Then, usually after just a few minutes that seem to go on forever, the child falls back asleep. Whether the parents can manage to get back to sleep is another question. And in the morning, they may be shocked to realize that their child has no idea what they are talking about.

Another common parasomnia, *nightmare*, is quite different from night terrors. As most people know very well, nightmares are terrifying dreams. They usually happen during the second half of the night, when REM dreaming is more frequent. Children who have had a nightmare wake up all the way, scared and upset. They seek to be comforted and may be frightened of going back to sleep, in case the nightmare returns. Both then and in the morning, they can usually recall details of the dream that scared them so. Most children have an occasional nightmare, and that should not be a cause for worry. If nightmares become frequent, however, and especially if the child is disturbed by daytime fears as well, parents should think about getting professional help with the problem.

Promoting Healthier Sleep

Getting an adequate amount of healthy sleep is as critical to children's development as having a sound, nutritious diet, and in today's 24/7 fast-food world, it is just as endangered. Too many adults think that getting by with less sleep is an accomplishment to be proud of. Children notice this and adopt the attitude for themselves. Even if they didn't, their lives are filled with noise, artificial light, and the constant flickering lure of electronic media. They do not find it easy to screen out the distractions and retreat into slumber. Yet it is urgent that they do so.

Sleep does much more than simply give the body time to recover from a day's activities. The benefits range across practically every area of development, from physical and cognitive to emotional and social. Children who get enough sleep are happier and easier to get along with. They focus their attention better, which helps them cope with school. They are less at risk for growth problems and obesity. And recent research suggests that adequate sleep strengthens the immune system, helping to ward off disease.

What do we mean by "adequate sleep"? Of course it varies from one child to another, and of course it changes as the child grows older. It may even change with the time of year and the school calendar. However, here are some rough guidelines:

- *Newborns*—16–18 hours per day
- *1 month–1 year old*—14–15 hours per day
- *1–3 years old*—12–14 hours per day
- *3–6 years old*—11–13 hours per day
- *6–12 years old*—10–11 hours per day

We should stress that this list does not tell you how many hours children actually sleep. Instead, it is a set of recommendations for how many hours they *should* sleep. And as the car commercials say about gasoline mileage, your results may differ.

Parents can do a great deal to promote healthier sleep in their children. Many of the steps they can take are common sense. We mention them anyway, because common sense so often gets forgotten or overlooked.

- A child's bedroom should be a cool, quiet place for relaxation and sleep, not for e-mailing and watching TV. And certainly *not* a place the child gets sent as punishment—that is a recipe for creating sleep difficulties.
- Set consistent bedtimes and wake times, even on weekends. A morning of "sleeping in" interferes with the normal daily buildup of sleep pressure and makes it harder to get to sleep at the proper time that evening.
- During the day, cut down on chocolate, cola drinks, and other sources of caffeine. Caffeine not only makes it harder to fall asleep, it leads to less sleep and lighter, less satisfying sleep. And it can go on affecting kids for eight hours or more.
- Make evenings relaxed. That means reining in active play, razzle-dazzle video games, and overstimulating TV shows as bedtime approaches.
- Create a calm, pleasant bedtime routine, such as a bath and a story before lights-out. Bedtime rituals are not just for preschool kids; older kids benefit from a predictable quiet time as well.
- The inner clock is very sensitive to light input, especially light in the blue part of the spectrum. Limit evening

exposure to TV and computer monitors. If the child is unhappy sleeping in a totally darkened room, a dim amber night-light can offer comfort without upsetting the sleep cycle (see *Resources for Follow-up*, p. 301).

Children and Mood Problems

Becoming depressed is not just a problem for teens and adults. Children, and even toddlers, can develop symptoms of depression. For some, these symptoms are long-lasting and severe enough to lead to a diagnosis of major depressive disorder. For many more, their condition never reaches the point of being noticed, diagnosed, and treated. Instead, parents see it—perhaps optimistically—as simply something their child is going through and will grow out of.

Unlike adolescents and adults, children are not likely to complain about being depressed, because they do not *know* they are depressed. They may think that their pervasive sadness, tiredness, and lack of interest, or the way they feel short-tempered much of the time, is just the way they are meant to feel. Parents can also easily miss, or misinterpret, their condition. Among the most common signs of depression in children are irritability and bad temper. But parents may think their child is simply mad about something and not see this as a possible symptom of depression. In any case, irritability is also seen in anxiety, learning, and attention deficit hyperactivity disorders, which are far more common in children than depression itself. Depressed children almost always have sleep problems, too, but so do many children who are *not* depressed.

Compared to the amount of clinical research on adolescent and adult depression, little attention has been given to prepubertal children. Even less has been done on the ways childhood depression may be linked to problems with the inner clock. Even so, we do know some things.

According to family research, the children of depressed parents are much more likely to develop a depressive disorder than those with non-depressed parents. Why this is so is still not clear. It may reflect an inherited predisposition, or it may be that depressed parents generate a family atmosphere that makes their children more susceptible to mood problems. Additionally, children who form weak attachments as babies—perhaps because the mother is going through postpartum depression or because the parents or caregivers have their own problems with attachment—are more prone to anxiety and mood problems.

Another characteristic of the depressed child—and depressed teens and adults as well—is a mind-set called "pessimistic attributional bias." This means that the child is likely to accept *responsibility* for negative events. For example, if the parents argue, children may assume that they themselves are to blame. This mind-set may come first, and trigger the depression, or it may be produced by the depression. In either case, it can go on shaping the child's thoughts after the depression passes, making a relapse more likely.

Treating Childhood Depression

Most often, childhood depression is treated using cognitive behavioral and interpersonal therapy. Sometimes antidepressant drugs are used, either alone or as an addition to therapy. Giving these medications to children is very controversial, although recent clinical trials of Prozac, for those eight and above, and Lexapro, for those twelve and above, indicate that they do work to some extent. We advise parents to give serious consideration to the risks and benefits before accepting antidepressants, especially since counseling and therapy are often effective in relieving childhood depression.

One promising intervention pairs parents and children in a guided form of play therapy that focuses on emotional development. The child learns how to recognize different emotions and take control over intense episodes. In initial study results, this technique benefits both the child and the parent. The child's mood improves, while anxiety, hyperactivity, conduct problems, hostility, and inattention are reduced. The parents, in turn, experience reduced depression and parenting stress.

What about chronotherapy for young people? We can gain some inspiration from a National Institute of Mental Health (NIMH) and Harvard University study of children who became depressed in the winter. Like nonseasonal depression in children, childhood SAD shows itself in irritability, difficulty concentrating, and problems at school—as well as in low mood states. And it can easily be confused with attention deficit hyperactivity disorder because of the overlapping symptoms. In the case of SAD, however, the symptoms vanish in the spring and summer.

In this clinical trial, one group of children—some as young as seven, but extending into the teenage years—had dawn simulators installed in their bedrooms. They also used a bright light box later in the day. A second group wore clear "placebo" goggles and received very dim dawn simulation while in bed. Partway through the season the treatments were switched between the two groups. About three-quarters of the kids showed major improvement when they were getting the active treatment, while only one-quarter improved during the placebo phase. This is a big difference for *any* treatment trial of clinical depression.

At our Columbia clinic, we have used bedroom dawn simulation with children as young as six, and bright light therapy starting in the early teens. These young patients have included both seasonal and nonseasonal types who had problems waking up for the school day. The conclusion we reached is that both dawn simulation and light therapy *work*, and they can be combined for extra effect—as they were in the NIMH/Harvard study.

Dawn simulation is the least intrusive method. The child doesn't have to do anything, because the lights start coming on automatically toward the end of sleep. It doesn't take time away from the child's day, and it's fun. (For details about dawn simulation, take a look at Chapter 7.)

In contrast, bright light therapy means a daily time commitment on a standard schedule. Some kids resist complying, even with—and maybe because of—parental insistence. Others make a game of it and do beautifully, but these have been few. Still others find the procedure embarrassing and hide their light boxes when friends come over. Finally, a small number—especially boys who have gone through puberty in the last year or two—experience a side effect we have not seen at other ages: heart pounding, sweating, and nausea, which prevents continuation of treatment. In general, we do not recommend using bright light therapy before the age of fourteen.

Whatever approach is chosen, childhood depression should not be ignored or dismissed as unimportant. Children who go through one bout of depression are miserable at the time, and they are likely to go through more of them—and more ominous ones—as adolescents and adults.

13. The Challenges of Adolescence

Melissa B. is fourteen and in middle school. Whether or not she can stay in middle school is becoming an issue. At least once a week she oversleeps and misses the school bus in spite of two alarms and a clock radio that blares rock music each morning. Even on days when she does make it to school on time, her teachers say she is no more than half-awake until afternoon. The kids still tease her about the day she started snoring during a discussion of *A Wrinkle in Time* in English class. And the worst part was that she loved the book and had lots to say about it, but she never got a chance to speak.

Melissa's parents don't know what to do. Her grades are slipping and the school has sent home warning letters about lateness. They let Melissa sleep in on weekends, but it doesn't seem to help. Neither does enforcing rules about bedtime. They have considered taking her to the doctor for sleeping pills, but are worried about starting her on them so young.

Upset by their nagging, Melissa says she just wants them to leave her alone.

Melissa has a point.

Sleep deprivation is an epidemic among adolescents. Anywhere from one-third to three-quarters of teens suffer from some kind of sleep disturbance. In one survey of three thousand high school students, 85 percent had trouble waking up for school and tended to fall asleep during their first-period class. And no wonder. Contrary to what both parents and teens often assume, adolescents need a good deal more sleep than adults, as much as 9.2 hours a night. Yet most of them don't get it.

Jonathan's Story—Part I

I knew something was wrong," Jonathan D. recalls. "I was waking up at 5 PM and going to sleep at like 5 AM. My grades were bad. I was falling asleep in class. I used to wake up at 4 PM or 5 PM, just as I was coming off the bus from school. My mother said I was sleepwalking through life."

Jonathan is fifteen and lives in suburban Louisville. He likes playing the guitar and thinks about performing with a band when he's older. He also likes playing games with his friends, but that hasn't always been so much fun. "I was so tired and stressed out. I would snap at them."

According to Jonathan's mother, his difficulties sleeping go all the way back to babyhood.

"He didn't seem to be able to sleep without my being with him as a soothing presence," she explains. "People told me to try this method where the kid is not supposed to be rewarded with a lot of attention, so they learn to soothe themselves. But Jonathan just felt alone. He wanted someone to come in, and having him cry for an hour was terrible. I felt I was abandoning my son and torturing him. Finally I moved a mattress, put it on the floor because I was just so tired. Or I would sit in the rocking chair. This went on from birth until he was about twelve, and I was still in the rocking chair for at least an hour."

Mornings were the worst, according to Jonathan. "It was horrible! I couldn't wake up. Dad said it was my personality. He said, 'You don't *want* to get up!' "

"To get him out of bed was almost impossible," his mother recalls. "A few times I just said, 'I'm sorry, I have to go,' and I left him at home. I was late for work and his sister was late for school so often. When he was little, the only way I could get him up was to pick him up and put his feet on the cold bathroom floor."

Jonathan's sleep pattern affected him at school as well. As his mother says, "He'd be falling asleep in class. How interested does a student look if he's falling asleep? And the teachers would wonder if he's doing drugs, or if he had family problems.

"It was never identified as a sleep issue," his mother continues. "They said that maybe it was obsessive compulsive disorder, maybe it was ADHD. One year, his grandparents died and I separated from his dad, so his problems were blamed on that. The summer when he was fourteen, I said, 'I'm curious what your body would do if it was left to itself.' The first week he slept from 1 AM to 1 PM. The second week, 2 AM to 2 PM, and it went on like that. How could he have a life with that schedule?"

Only about one in ten adolescents gets as much as eight and a half hours of sleep on school nights. Even more worrisome, one in four gets less than six and a half hours—that is over two and a half hours less than they need. It is not surprising, then, that sleep researcher James Maas has said that almost all teenagers become walking zombies. They do get more sleep on weekends, true—as much as two hours more a night. But this does not make up for the sleep lost on school nights. In fact, these irregular schedules simply make getting to sleep on school nights even *more* difficult.

What is going on? One of the major events at puberty is that the child's inner clock changes its settings. Suppose that before puberty it has been signaling that it is time to sleep at 9 PM and time to wake up at 7 AM. Now it shifts in the direction of signaling sleep at midnight or 1 AM and waking up at 9 or 10 AM. This shift happens at a younger age for girls, who go through puberty first, and happens earlier for teens who enter puberty sooner than their age mates. And this isn't just a computer-, TV-, and Twitter-driven American problem; it happens in the U.S., Europe, and Asia, and across cultures that range from pre-industrial to postmodern.

If the schedules that society imposes on adolescents were to shift in line with the changes in their inner clocks, sleep deprivation would not be a problem. They could go to bed after midnight and get up in plenty of time for lunch. Instead, they are expected to wake up at the same time or even earlier than younger children to get to school. The problem is easy to see: *late bedtime + early wake-up time = not enough sleep*. As simple as that equation looks, the consequences are not trivial.

Sleep deprivation in teens is linked to problems that go well beyond dozing off in morning classes. These can include poor school performance, negative moods such as anger and sadness, general "moodiness," problems with emotional control, conflicts with teachers, weight gain, increased aggressiveness, and problems staying focused on a task. In other words, sleep deprivation can cause major social, psychological, and physical problems for teens.

Even more alarming, drowsiness and attention lapses are major contributors to traffic crashes, the number one cause of death among teens. In a recent poll, 27 percent of adolescents who drive said they had had an accident or near-accident due to drowsiness, and 5 percent were willing to admit that they had literally fallen asleep at the wheel at least once.

What about the neurological consequences of sleep deprivation? Adolescence is a crucial period for the development of the prefrontal cortex—that part of the brain necessary for planning, self-control, and creativity. Loss of sleep interferes with this development. And recent research indicates that with too little sleep there are changes in areas of the brain involved in reward, sensation seeking, and risk-taking. When teens have to decide whether to use drugs, drive recklessly, or have unprotected sex, it is their brain's reward system that weighs the risks and benefits. Lack of sleep can mean that your teen is trying to make life-changing decisions while impaired to a degree that we can't even measure yet.

Why are adolescents so likely to develop sleep problems? Some of the reasons we will look at are biological, some are psychological, and still others are social and cultural. Taken together, they create a downward spiral in which each element works on and magnifies others. The particular pattern differs from one teen to another, but practically all teens are affected to some degree by all these factors.

Biological Changes

On the most basic level, adolescence is a biological event. It marks the physical and sexual transition from childhood to adulthood, and it happens even in species that have never heard of teenagers. During puberty, the body's endocrine system starts producing dramatically higher levels of sex hormones—estrogens and androgens. The most noticeable result is the appearance of secondary sex characteristics, such as breast development in girls and pubic hair in both boys and girls. But these hormones also have a marked impact on the brain and central nervous system. Although there is much about this process we still don't understand, both the circadian clock, in the hypothalamus, and the pineal gland, which produces melatonin, are affected. And these are crucial in determining the timing of someone's sleep/wake cycle. (For a fuller explanation about how the sleep/wake cycle is organized, take a look at Chapter 3.)

Scientists have pointed to several developments that are linked to puberty and that combine to push the sleep cycle later.

- The daily rise and fall in melatonin release comes later and later as children move through puberty, so they don't become sleepy as early in the evening and don't become alert as early in the morning.
- The adolescent's internal clock slows down. This also tends to push sleep time later.
- The amount of melatonin produced drops dramatically. This is part of the normal pubertal process, and the decline will continue across adulthood.
- The pressure to sleep during the waking hours accumulates more slowly in teens than in children, and teens can also resist it better. So they don't become sleepy as soon, and even when they do, they can still manage to keep themselves awake.
- Puberty also results in *lowered* sensitivity to the effect of morning light, which pushes sleep time earlier, and *heightened* sensitivity to the effect of evening light, which pushes sleep time later.

These biological changes affect all adolescents to some extent. Other factors are more specific to particular cultures, families, and individual teens.

Psychological Factors

"Please, can I stay up? Just a little longer? I'm not sleepy!"

Are there any parents who have not heard this plea? In some families, bedtime is a dreaded daily battle—in others, it is a familiar and comforting routine. Yet even kids who get into their pajamas without being reminded and are asleep moments after their heads hit the pillow often see going to bed as missing out on something fascinating—grown-up events. In the child's view—one often shared by adults—growing up is signaled by such events as moving to a higher grade, being able to do more things, and having a later bedtime. Thus, the familiar protest, "I don't want to go to bed yet. I'm not a baby anymore!"

To complicate the issue, parents often let children stay up later as a special reward or to mark special occasions. For the child, this makes it even more obvious that going to bed later is a privilege.

Which brings us to adolescents, who according to one definition are no longer children but not yet adults. They are drawn to activities that make it clear they belong to the high-status group—grown-ups—and *not* the low-status group, little kids. Going to bed later, or not at all, is definitely one of them. Staying up late gives you bragging rights, whether you did it to instant-message your ten closest friends, watch a movie, study for a test, or play your favorite video game.

Adolescents are also deeply concerned with the shifting boundaries of their independence. Do they have to do as they're told, or can they decide for themselves? And on which issues specifically? Teens generally agree that parents have the right to make decisions about issues of safety, but they battle fiercely for the right to make their own decisions about personal issues such as friends, music tastes, and clothing. Where does bedtime fall in this spectrum?

Surveys show what most of us would guess. Among ten- and eleven-year-olds, about half say their parents set their bedtimes, at least on school nights. By age twelve and thirteen, the proportion has dropped to one-quarter. And practically no high school students admit to such a childish situation. But how do you go about *demonstrating* that you make your own decision about bedtime—instead of following your parents' orders?

By turning in at eight? Possibly, but a much more obvious step is to make a point of staying up later than your parents think you should.

Staying up late is not simply aimed at parental rules, however. Even adolescents whose parents brag about their model kids can stumble into the late-night trap because there is so much they must do and not nearly enough time to do it in. Burning the candle at both ends is more than a vivid image: It is a practical necessity.

Teens may be given too much homework, which can take up several hours every night. They also have to cope with a widening array of non-academic obligations. For some, these might include sports practice, rehearsals for chorus and the school play, after-school courses to help them get into college, household chores, required community service, and part-time jobs. Others, from less advantaged homes, are burdened with shopping, cooking, household chores, and acting as caregivers to younger brothers and sisters. And in both cases, what about the time spent daydreaming about that boy or girl in history class whom you've never spoken to but just know is the love of your life? However these elements add up for the teens in your house, chances are they're subtracting from their sleep time.

Friends, Home, and School

Something else everyone knows is that during adolescence the peer group takes on major importance. American teens spend two to four times more waking hours interacting with other teens than they do with their parents or siblings. Much of this interaction takes place in school, of course, but it continues after school, on weekends, and during vacations, too. With evening comes a constant barrage of tweets, chats, texts, IMs, e-mail exchanges, and even old-fashioned telephone calls. This has a direct and obvious impact on sleeping habits. Although no one has kept track, it is a safe bet that one of the more frequent text messages teens send or receive is *R U up?*

In the early years of the electronic age, the television and computer were ensconced in the living room or family room, where everyone could use them. Today, practically all American teens have electronic gadgets in their bedrooms: music players, TVs, video game consoles, telephones, and of course, computers, almost all of which have Web access. Young teens have an average of two devices, and older teens an average of four. The results are pretty clear: They can be in touch 24/7 if they want, and sometimes it seems as if that's exactly what they're shooting for.

The more media devices teens have in the bedroom, the more likely they are to report that they:

- go to bed later
- get less sleep, on both school nights and weekends
- have more restless sleep
- feel more tired during the day
- fall asleep more often in school
- drink more caffeinated beverages during the day

Those late evenings spent staring at a television or computer screen—or for serious students, at the brightly lit pages of a textbook—have a more subtle impact as well. Exposure to bright light in the evening hours, before the sleep cycle begins, has the effect of pushing both the circadian clock and the release of melatonin later. (For a fuller explanation about the effects of evening light, take a look at Chapter 1.) This in turn feeds into the cycle by allowing the teen to stay at the computer later before feeling too sleepy to go on. There is a price to pay, of course, but the bill won't fall due until the next morning at 6 AM, when the alarm goes off.

Why so early? School, of course. Until fairly recently, the usual start time for school was 8 AM. Today, however, many American adolescents have to be in school as early as 7:15 AM. This means catching the school bus at 6:45 AM or even earlier. One explanation for this change is economic. If the start times of elementary, middle, and high school are staggered, the same school buses and drivers can make three separate runs. Only one-third as many buses are needed, which cuts costs quite a lot. However, it also means that during the winter months, at least one age group is forced to go to school or come home from school in the dark. Which? Faced with the image of eight-year-olds huddled by the roadside in early-morning darkness, school administrators generally decide that adolescents are the best equipped to get up earlier.

A further complication is that about 50 percent of teens who are in school also have jobs, and half of those who work put in twenty hours a week or more. Add that to the thirty or thirty-five hours a week spent in school, throw in some time for meals, friends, getting to and from school and work—and an occasional conversation with parents—and it is clear that something has to give. What gives is sleep.

Our Advice for Parents

"You don't understand!"

If you are like most parents of adolescents, you have heard this plaintive cry more often than you would like. In calmer moments, you have probably admitted to yourself that at least sometimes, you truly don't understand. You may know perfectly well that your teen is not getting enough sleep, resists going to bed on time, drags around for half the day and behaves like a zombie on weekends, but you don't know why. Adolescent rebelliousness? Raging hormones? A basic character flaw? Your fault?

Generations of American parents have quoted Ben Franklin: "Early to bed and early to rise makes a man healthy, wealthy, and wise." And generations of teens have rolled their eyes and sighed, *"Puhleeze!"* With reason, as it turns out. Franklin's advice may work for adults, but most teens are dealing with too many biological and social pressures that push them in the opposite direction. It does not help if you constantly criticize them for something they have so little control over.

Just as it is always a good idea for parents to avoid blaming children for something that is not their fault, it is also a good idea for them to help their children avoid blaming themselves. As a parent who wants to help a teen with sleep difficulties, you are already making the effort to educate yourself about the causes of those problems. The next step—equally crucial—is to talk about the problem with your teen. Make sure that both you and they understand the extent of their difficulties.

First, ask your teen some basic questions:

- On school nights, do you fall asleep later than you think you should?
- Do you have to use an alarm clock—or two of them—to wake up on time?
- Do you sleep through the alarm or keep hitting the snooze button?
- Do you need to catch up on sleep on weekends by getting ten hours or more?

Then explore together how these sleep habits are affecting their life:

- Do you spend much of the morning at school dragging around half-asleep?
- Do you sometimes nod off in early classes?
- Do you feel you need to rely on energy drinks (or pills) to get anything done?
- Do you feel as if your brain finally starts working in the afternoon?

- Do you realize how often we're getting into arguments over your sleep habits?

The more your teen answers yes to these questions, the clearer it is that problems with sleep are having a negative impact on his or her life. To get a more detailed understanding about what's going on, teens should take the online chronotype questionnaire (see *Resources for Follow-up*, p. 301). The personalized feedback will help them find out how much of a lark or owl they are compared with other people. When does their internal clock think they should be going to sleep? Is that when they are *actually* getting to sleep?

If you suspect your teen is showing signs of depression, there is the confidential online self-assessment of depression severity. This will help them to face the next steps and, if indicated by the questionnaire's personalized feedback, to seek help. Antidepressant drugs may not be the answer. In a recent drug-company study of more than three hundred children and adolescents with depression, Prozac seemed to help the kids, but not the adolescents! Our interpretation is that adolescent depression is closely associated with the inner-clock shift later that comes with puberty, and Prozac is not a chronobiologically relevant drug. What's needed is to shift the clock earlier, in sync with sleep—in other words, chronotherapy.

Now that you and your teen see how many problems these sleeping patterns are creating, what next? First, it is crucial to understand that having these difficulties does not mean your son or daughter is lazy or crazy. And the fact that they (and you) are finding these problems hard to deal with does *not* mean they lack willpower! This is what being an adolescent is all about—changes. The physical, hormonal, and neurological changes that began with puberty push their internal clock in the direction of a later daily rhythm. It happened to you when you were their age, it happened to your friends and classmates, it happened to teens on the other side of the world, then *and* now. And today's 24/7/365 culture has dramatically worsened the problem. It is a universal fact of adolescence, just as much as getting taller or moving toward sexual maturity, and it has to be dealt with by you and by your child.

However, it is equally important to recognize that biology is *not* destiny. Adolescents do not have to put up with sleep or mood or energy problems just because these problems have physiological causes. That would be like saying, because some kids are physiologically nearsighted, they should resign themselves to never being able to see the blackboard. Of course not! Instead, they are given an eye exam and get glasses or contacts to correct the problem.

Of course, as a caring parent, you will do what you can to help, but in the end teens need to take responsibility for dealing with their sleep issues themselves. The more they understand the sources and consequences of their sleep problems, the more they are likely to want to change and believe that change is something they can really accomplish. If they feel the necessary changes are being imposed, however, you are more likely to find resistance instead of cooperation. Discuss what you have learned. Suggest that they read this chapter, especially the section below, "Our Advice to Teens." Then propose arriving at an agreement that lays out the steps they intend to take to deal with their sleep issues. Adolescents are often hypersensitive about being told what to do, but they can also appreciate the concern and support implied by reasonable rules. Three aspects of your teen's daily life call particularly for your input:

- *Bedtime.* Always staying up late when you know you have to be up early the next morning is *not* a badge of adulthood—it's a sign of a problem that needs to be dealt with. Discuss *with* your teen what a reasonable bedtime would be, and then help him or her lay out the necessary steps to reach that goal and make it a habit.

- *Electronic media.* The latest research makes it very clear that looking at a TV or computer screen in the evening pushes the body's systems toward a later bedtime. Encourage teens to set their own time limits on exposure. Remind them how much better they have felt after a good night's sleep. Setting time limits on exposure might help tremendously—although it may not meet with much cooperation from older teens.
- *Employment.* Full-time students who also work at after-school jobs are a peculiarly American phenomenon. Quite simply, our kids are doing too much for the number of hours in the week, and what usually suffers is getting enough sleep. If your teen is asking permission to get a part-time job, give some serious thought *together* to the advantages and disadvantages. Obviously, this issue cannot be resolved if a job is financially necessary. Still, think seriously about time spent working compared with the number of hours of lost sleep.

Making a Family Contract

Many families find it helpful when dealing with difficult issues such as schoolwork, chores, and curfews to negotiate a family contract between parents and their teens. This sets forth in writing what behaviors the parents expect, what rules and regulations the teen agrees to follow, and what the consequences will be for either meeting or not meeting these expectations. If you and your teen decide together to adopt this approach in dealing with sleep difficulties, we suggest the following steps:

- Teens take a serious look at the next section of this chapter, headed "Our Advice to Teens." They write down a list of the specific changes that sound reasonable and helpful.

They also indicate what they think would be appropriate consequences—both positive and negative, and list the kinds of support they would like parents and siblings to commit to.

- Meanwhile, you do the same. Think about such questions as: *What is a reasonable bedtime on school nights? On weekends? How much television and computer time, if any, during the evening?* and so on.
- Hold a family discussion to identify no more than five changes that the teen makes a commitment to carry out. Be specific about these: not *I'll try to get to bed earlier on school nights,* but rather *On school nights I will be in bed with lights out by midnight.*
- For each change, list specific payoffs for meeting or not meeting the goal. For example: *If I am in bed on time for an entire week, I can choose a weekend trip or activity. For each ten minutes I stay up past the stated time, I lose thirty minutes of television or computer time the next day.*
- List the ways in which other members of the family will cooperate, such as: *We will gradually dim the lights and reduce the noise in the living room beginning at 10 PM, and not watch television in the living room after eleven.*
- Agree to revisit, revise, and renegotiate the contract after, say, thirty days. It is important to talk about what has worked and what hasn't, and make any needed changes. It is just as important not to try to accomplish everything at once. Small successful steps are much more valuable than ambitious programs that fall flat.

It is important to remember that this is a *family* contract. It is meant to help solve a problem that is as much a family problem as your teen's problem. By taking everyone's ideas and concerns into consideration, and getting everyone's agreement to cooperate, you will make it much more likely that your teen will succeed in dealing with his or her sleep problem.

Our Advice to Teens

Do you wonder if difficulties with sleeping, mood, and energy levels are getting in the way of having the kind of daily life you want? If so, you shouldn't be surprised. These problems are common, and can have many causes, for teens as well as for adults. Sometimes they become so severe that they need to be treated by a professional who is skilled in using chronotherapy or psychotherapy. On the other hand, you may well find relief just by taking some simple steps on your own.

You *can* decide to improve your sleep, mood, and energy. Learn the reasons for the problems and how they can be solved. Take concrete steps that fit your goals, personality, and any external issues. Then resolve to carry out your plan, since no one else is going to be responsible. There will probably be setbacks. It won't always be easy to keep your long-range goals in mind when a more immediate temptation comes up. But if you can stay with the program you set yourself, you are sure to see encouraging changes.

Some of our pointers may sound totally obvious, but please don't brush them aside. Instead, ask yourself: *If they are so obvious, why haven't I been following them?*

- *Get more sleep.* There is nothing you can do about the time you have to get up on school days. So any changes have to take place at the other end, with bedtime. It will work best to change gradually. Let's say you now go to bed at 12:30 AM and fall asleep by 12:45 AM. Make a commitment to yourself to be in bed by 12:15. The first few nights you may stay awake until 12:45, but don't give up. Soon you'll fall asleep by 12:30. Once that's done, move your bedtime back to midnight. Keep the process going until you reach the bedtime you need to get the amount of sleep you need.

- *Get more regular sleep.* Try to get to bed at the same time on weekends as during the week and to get up weekend mornings after no more than eight or nine hours of sleep. If you are getting less sleep than you need on school nights—as is true for practically all teens—the pressure to sleep builds up during the week. The urge to catch up on the weekend is very hard to resist. Plus, of course, Friday and Saturday evenings are prime time in your social life. However, those hours sleeping in on Saturday and Sunday morning do nothing to help you during the week. In fact, they make it harder to keep to a better schedule once Monday rolls around. Sleeping in also allows your internal clock to drift later, because you are waiting too long before you get to see daylight. This is what makes Monday morning the most dreaded time of the week.
- *Get more morning light.* Bright light plays a critical role in keeping our inner clock in sync. It also has a powerful effect on our mood and energy. The simplest way of taking advantage of this process is to go outside at the end of your normal sleep cycle and spend at least ten or fifteen minutes in full sunlight—assuming you've had a decent night's sleep. If that means while you're waiting for the morning school bus, stake out a sunny spot and face east. (Clearly, this won't work if you've had only a few hours of sleep, or if you're outside waiting for the bus while it's still dark.)

- *Get more exercise, but not in the evening.* Exercise in the morning helps to synchronize the internal clock, making it easier to get to sleep earlier at night, but late night exercise has a risk of making it harder to get to sleep.
- *Cut down on caffeine, alcohol, and nicotine.* Scientific studies—like most people's personal experiences—make it very clear that caffeine, alcohol, and nicotine have bad effects on sleep. By limiting your consumption during the day and cutting it out altogether during the evening hours, you can improve your sleep *and* your general health. And do we need to point out that alcohol and nicotine consumption is not just unwise but illegal for most teens?
- *Cut down on evening light.* Exposure to bright light during the evening pushes the inner clock later, making it harder to get to sleep on time and to wake up the next morning. The TV, computer screen, or brightly lit book is making your sleep problem worse. Does this mean you should sit in the dark doing nothing until bedtime? Of course not, but try to reduce bright light as the evening goes on. Put your room light and desk lamp on dimmers, so you can see everything easily while avoiding overstimulation. Avoid using bright white fluorescent lamps, including the energy-saver type. One step that can help a lot is to download an application called *f.lux*, which gradually changes the color balance of your computer screen during the evening, and reduces the short wavelengths of light that have the biggest negative impact on your inner clock (see *Resources for Follow-up*, p. 301). It's pretty cool.
- *Unwind with a pre-sleep routine.* You will find it helps to create a regular transition between the stresses of the day and going to sleep—an appealing routine that puts you in a

calmer frame of mind. What this looks like is very much up to you. Lots of people find that a cup of warm milk or uncaffeinated herb tea helps. Others swear by meditating, taking a warm bath, or listening to soothing music. What's important is that it work for you and that you make it such a regular habit that you don't even have to think about it.

- *Get a little help from your friends.* Tell your best friends that you are going to deal with your sleep problems. By letting them know, you are making a sort of public statement, which is usually easier to keep to than private decisions. Ask them to help by not texting or calling you late in the evening. Just for insurance, turn off your cell phone audio alerts for the night. Your friends may tease you, but they are probably having the same problem. Why not rope them in, and use the buddy system?

If you are feeling depressed, anxious, or both—and this is interfering with your sleep, schoolwork, home life, social life, or even your outlook on life—don't keep it to yourself. There are many possible causes as well as ways to deal with the problem so your mood improves rather than getting worse. Take a deep breath, and don't delay sharing this with your parents, a friendly teacher, guidance counselor, or doctor. If you are in the U.S. and don't have someone to talk to, go to crisischat.org or call 1-800-273-TALK, where everything is private. Outside the U.S., go to befrienders.org to find the right person.

And don't forget to quickly help your friends if they tell you about such feelings. Start by asking your parents what to do to help. Here's an example. (Use your own words, of course.) "Mom, I've got to tell you something about Paul. He's hardly been talking to me for two weeks. It just doesn't make sense. He seems so remote and sad and wrapped up in himself. It's making me worried. We didn't have a fight or anything. Should we say something about it?"

What Communities Can Do

The widespread sleep problems among teens are not just a matter of biological changes or the intrusion of late-night social contact. Local communities must accept a share of the responsibility as well. Practically all middle schools and high schools force teens to get up much earlier than their natural wake-up time. The results include wasted first-period classes—and second, and third—chronically tired students, and soaring sales of highly caffeinated energy drinks, not to mention stimulant-pill abuse.

Just stating the problem seems to point to an obvious solution. Why not start school later? A few communities have indeed tried it. In 1997, Minneapolis shifted the start times of all its public schools so that high school started at 8:40 AM and middle school at 9:10 AM. Attendance went up, lateness went down, and fewer students fell asleep in class. Students also reported fewer feelings of depression.

More recently, researchers in Rhode Island compared students before and after their high school start time was moved from 8 AM to 8:30 AM. This half-hour delay led to a forty-five-minute increase in sleep time, better attendance, and a reduction in fatigue, sleepiness, and depressed mood. The students were also more motivated to take part in sports and social activities and less likely to go to the health center for fatigue-related problems.

Why, then, aren't more communities adjusting their school schedules to meet the needs of their teens? The proposal to change school start times often arouses strong emotional reactions. Even though high schools start much earlier today than they did a generation or two ago, the early times seem to have taken on the firmness of long-established tradition. They have become an integral part of the daily norm. Parents and teachers have come to expect and rely on them. And more than one school official has found that trying to change to later start times caused controversy about and challenges to their leadership.

Some doubters assume that if teens don't have to be in school until an hour later, they will simply stay up an hour later. Here, the research results are quite clear. In Minneapolis, students in the study got an average of an hour more sleep than similar students in schools with an early start time. The same was found in a comparison of early- and late-starting middle schools in a New England community.

Of course, delaying the start of school is not going to be a cure-all for adolescent sleep problems. Even with an hour more, most are still getting less sleep than needed on weeknights, and a lot are still going to bed later and sleeping longer on weekends. In addition, they may still be exposed to early morning light at a point in their circadian rhythm that tends to push their sleep cycle later, rather than earlier. However, the benefits of starting school later, at an hour that better fits the sleep cycles of teens, definitely outweigh the downsides of making the change. As a parent, citizen, and voter, you can help make sure your local school officials are aware of this important issue. They should know that you and others in the community want them to put the health and alertness of students ahead of bureaucratic priorities.

Schools can be helpful in other ways as well. Most teens take classes at some point that address health issues important to adolescents. Sleep certainly belongs in that category. Do the health classes in your local schools deal with sleep problems? Do the teachers explain why these problems become increasingly widespread during adolescence, and how teens can try to deal with them? If not, the reason may be that teachers and administrators simply don't realize how much these difficulties affect their students and hurt the educational process. Educate the educators, and press them to educate themselves. Propose that middle schools and high schools hold teacher workshops on sleep issues. Chances are, teachers who see students in their morning classes nodding off will be very relieved to learn that it is not entirely their fault!

Jonathan's Story—Part II

In April 2010, Michael Terman wrote an article about sleep problems and chronotherapy for the *New York Times*. Jonathan's mother happened to notice it. "As I read it, I said, 'That's my boy he's talking about!' I e-mailed Michael right away."

"Mom said we have to do it, give it a try," Jonathan says. "My grades were bad, I was falling asleep in class. I figured there was nothing to lose. I didn't know what to expect, just that it used lights and it shifts your circadian rhythm back to normal hours."

In mid-August, Jonathan and his mother flew to New York City. They were both interviewed at Columbia University Medical Center by psychiatrist Gregory Sullivan at the Center for Light Treatment and Biological Rhythms. In his notes, Sullivan described Jonathan as fidgety but well groomed, with normal speech and "okay" mood. He concluded that Jonathan's medical history was consistent with delayed sleep phase disorder, and endorsed treatment with bright light therapy and melatonin.

After intensive talks with Jonathan and his mother about his history and sleep pattern, Michael developed a treatment plan custom-tailored to Jonathan's situation.

"I'd read enough about the light box and the melatonin," Jonathan's mother says. "But I didn't expect such specific recommendations as Michael made to Jonathan. I also did not expect his rhythm to be so obviously distorted, or to be so much longer."

"Dr. T said I was physically in New York but my circadian rhythm was in California," Jonathan recalls with a laugh. "I remember the first time I did the lights. I had to do it at eleven AM for half an hour, and getting up at eleven was brutal, especially over the summer. I was going to bed just seven hours before that."

Jonathan's mother noticed a change right away. "We set up the lights, and the third day I wondered, 'Am I dreaming?' You know how you have a plant that's drooping—it's not dead, but it's really drooping. You give it water, and within an hour the leaves extend. It was like that with Jonathan, it was like someone watered him! He's nicer, more positive. Even his sister has to agree. He's not as irritable. Before, if she batted her eyelashes the wrong way, he'd get really upset."

Asked about the light treatment, Jonathan says, "The first couple of times it was hard, but it got easier. The time started being pushed back, and at the end of the summer it was being pushed back every morning by half an hour. That was difficult, but I did it."

"Jonathan's been completely cooperative," his mother points out. "He's fascinated by the treatment, and knowing it's not a prescription medication appealed to him a lot. He didn't want to visit another provider and be dependent on pills. My only concern is that I could imagine that if he wants to be normal, with a normal circadian rhythm, that he probably has to do this all his life. For example, in college, what will happen?"

After a moment, she adds, "I am so proud of his desire and ability to take ownership of this. It's nice to be able to talk to him about the fact that he had something that could be treated and that he's *not* a lazy, disobedient kid like some of his teachers thought."

If his teachers did think that, they don't any longer. "Things are great in school now," Jonathan says proudly. "I'm getting all A's and B's. I can stay awake in all my classes. My French teacher noticed it. They say this is the real me."

Asked if any of his friends have sleep issues, Jonathan laughs. "Sure, everybody! But I don't know anybody whose problems are as bad as mine."

What Treatment Can Do

Because the biological clock is set differently from one person to another, the sleep problems of individual teens vary substantially. In recommending solutions, it is a big mistake to treat anyone as an "average case." Lucky kids will find that something as simple as reducing evening room light will take care of their problem. Less lucky ones will find that no matter what they try, their energy and alertness are peaking at 2 AM, long after they should have fallen asleep.

Let's say that you've read the recommendations in this chapter. You've discussed them with your teen, perhaps negotiated a family contract, and watched your teen put in a sincere effort to achieve a healthier, more satisfying sleeping pattern. But after a few weeks, you realize that the needle hasn't moved off zero. The problem is not getting better. At this point, even if your child's difficulties don't seem that severe to you, it is time to consider getting outside help. Sleep disorders are a recognized medical condition, and there are specialized approaches to treatment that are both safe and effective.

The nicknames of larks, owls, and hummingbirds are convenient shorthand, but in the research world we also talk of moderate evening types as distinct from definite evening types. Beyond this range, there are distinct *night types* whose circadian clocks have slowed down to the point that sleep comes only in the wee hours, or not until morning approaches. The teen who has been fated to fall into this group can still achieve normal sleep, but not without professional supervision of chronotherapy. Self-treatment almost surely will fail.

If an adolescent is diagnosed with delayed sleep phase disorder, the clinician will set out to develop a treatment plan that is appropriate for the specific situation. The first step is to verify the patient's spontaneous sleep pattern. A teen who guesses that he or she would fall asleep at 3 AM without any restrictions may in fact not fall asleep until 5 AM if left free to do so. That would put the end of internal night at around 1 PM.

Two essential tools in treating the teen's condition are bright light and melatonin. The light treatment is scheduled for the wake-up end of the sleep cycle and microdose, controlled-release melatonin is taken about six hours before current bedtime. (For a fuller explanation about melatonin, take a look at Chapter 8.) The light and melatonin both act to shift the inner clock—and therefore the sleep cycle—earlier. After taking the day's melatonin, the teen needs to avoid daylight and bright artificial light—such as at the shopping mall—or else to wear special sunglasses that filter out short-wavelength light, as we also explain in Chapter 8.

Adjustment of the timing of chronotherapy requires the clinician's continual guidance. There are too many ongoing decisions to be made for the patient to self-treat or for parents to manage. The dose of light and melatonin may need to be adjusted. The decision to move each step earlier must be considered carefully, because moving too fast can cause an immediate relapse. Therefore, updated sleep logs need to be faxed or e-mailed to the clinic every few days, where they are reviewed in order to plan out the next few days.

A treatment plan of this sort calls for the teen's real commitment. It is too easy to sabotage chronotherapy by playing with (or "forgetting") the timing of melatonin and lights, or even not doing it while telling parents and clinician that everything is going well. The best time to begin the treatment is during a three-week winter or spring break or during summer vacation, when the teen can sleep without the constraints of the school day. It may take ten days to several weeks to shift the internal clock into sync with local time, which means it is not very practical to begin the procedure while school is in session. But provided the teen is motivated to see results, it almost always works.

And that is very good news for adolescents with serious sleep problems, as well as for their worried parents.

14.
In Later Years

It's 5 PM. The afternoon sunlight filters into the living room through the drawn lace curtains. Paul R., eighty, sits in his easy chair, leafing through a magazine. Every few minutes his eyes close and his head starts to fall forward. His wife used to urge him to be more active, but she has given up. At 6:30 PM, as usual, she puts dinner on the table and they eat silently. By 7 PM, they are back in the living room in front of the TV. The volume is set high, because Paul is having hearing problems. By now it is dark outside, but only one lamp is on in the room. It feels calmer that way, and it helps save on electricity.

Before ten minutes have passed, Paul starts to doze off. When a blasting commercial jars him awake, he mutters something nasty and drags himself off to bed. This has been the pattern of most evenings for the past year or more. Once in bed, he falls asleep right away and gets up only once, around 11 PM, to visit the bathroom. But at 2:30 AM, Paul wakes up with a start. He looks at the lit numbers on the bedside clock and sighs. He knows from repeated experience what this means. The night is over for him. Nothing he can do will let him get back to sleep. Silently, careful not to wake his wife, he gets up and shuffles into the living room. It is dark and quiet outside. The sun won't be up for another four hours, but Paul is once again awake and alert in *l'heure du loup*, the hour of the wolf.

Age and Poor Sleep

Nationwide polls indicate that some two-thirds of American seniors—about thirty-seven million people—suffer from frequent sleep problems. Of these, only a small fraction have had their sleep problems diagnosed and treated. The others may assume that having sleep problems is simply a natural, unavoidable aspect of getting older. Their family members and doctors probably agree. Until recently, so did most sleep scientists. But when researchers looked at the sleep patterns of healthy people across the life span, they found that most of the changes happen before the age of sixty. Compared to teens, healthy middle-aged and older adults sleep a little less and wake up a little more often, but they do not take any longer to fall asleep or, once asleep, to start dreaming.

Poor quality sleep is not something to be accepted or brushed aside at any age, but the consequences are particularly risky for the elderly. Among these are:

- Daytime sleepiness and fatigue, leading to a lack of exercise that in turn may accentuate weight problems, diabetes, and difficulty getting around.
- Cognitive confusion and problems with memory and concentration, which a senior may see as indicating something much more frightening than lack of sleep.

- Depression, which can both lead to sleep problems and be a result of them.
- A weakened immune system, which increases susceptibility to disease.
- Increased sensitivity to pain, which in turn makes sleep even more difficult.
- Greater danger of nighttime falls and accidents.

In addition, sleep problems complicate the treatment of other serious medical conditions in older people.

Drugs, Light, and the Eye

Two important causes of sleep problems in the elderly are other illnesses and the drugs that are used to treat them. Many prescription drugs have been linked to insomnia, either by themselves or in interaction with other drugs. When we consider that the average senior takes six to eight prescription drugs, the odds are high that at least one of these is having a negative impact on sleep. Other sources of sleep difficulties include apnea (brief, repeated breathing interruptions), restless leg syndrome, and bladder problems that lead to nighttime urination. All these are well known and often discussed. Much less widely recognized—even by sleep experts and the medical community—are the ways the circadian clock is implicated in the sleep problems of older people.

Bright light is the most intensively studied time cue for keeping the inner clock in sync with the day/night cycle. It is also crucial to maintaining optimal alertness during the day. The presence of light, and particularly of its blue component, is signaled to the inner clock through a dedicated neural pathway from special retinal cells that contain the photosensitive chemical *melanopsin*. (For a fuller explanation, take a look at Chapter 4.) But as we get older, what is bright outside does not necessarily translate to brightness in the eye and brain. Starting around age forty, the liquid media of the eyes begin to cloud. In effect, this puts a density filter between the external world and the retina, reducing the amount of light that gets through.

In later years, the lens of the eye also changes, turning yellowish. This yellow coloration has the greatest impact on short-wavelength blue light—exactly the part of the visual spectrum that most affects the circadian clock. This reduces the amount of short-wavelength (violet, indigo, blue) light that reaches the retina, causing not only color distortion but also reduced stimulation of melanopsin, the blue-sensitive circadian photopigment in the retina. The amount of blue light that reaches the retina is reduced to as little as 30 percent of normal because of this natural aging process. We are not aware that it is happening because it takes place so gradually during later adulthood. But if you have had cataract surgery and lens replacement, you were probably struck afterward by how vivid colors suddenly appeared. The reason was that your eyes were once again sensitive to the entire color spectrum.

If you are faced with a decision about lens replacement, there are two main choices for the implants: clear and yellow-tinted. Some doctors recommend clear replacement lenses, and patients relish the prospect of seeing full-bodied blue again. On the other hand, the natural yellowing of the lens may serve a protective function in later years. Research suggests that excessive short-wavelength exposure—whether pure blue or couched in white light—accelerates the course of age-related macular degeneration (ARMD), the progressive blinding disease. The solution to this puzzle lies in dividing the short-wavelength range into two parts: "short-short," from 400–450 nanometers, and "long-short," from 450–500 nanometers. "Short-short" is the region specifically linked to ARMD. "Long-short," on the other hand, enables us to see clear blue and also optimally stimulates the circadian timing system. Yellow-tinted lens replacements of the Alcon type selectively block the problematic "short-short" wavelengths while transmitting the "long-short," for triple benefit: improved color perception, improved inner clock function, and retinal safety.

Fixing Light Loss

How can the problem of light loss in aging eyes be approached without cataract surgery? It might seem that the simplest answer would be to use light sources with a higher concentration of blue wavelengths. However, that would create other problems. The high-energy photons of blue light scatter when they hit the cloudy ocular media and the lens (with or without cataracts). The result is unpleasant levels of glare that cause squinting, eyestrain, and lowered visual sharpness. This may be why so many people dislike sharing the road with those blue-ish "high energy discharge" headlights found on some luxury cars.

Boosting levels of blue light is probably unwise for another reason, too. As we pointed out earlier, exposure to blue light in the "short-short" part of the spectrum can lead to worsening of ARMD. Macular degeneration is already on a steep rise, from around ten million people in the U.S. today to a projected eighteen million people by 2050. We certainly do not want to suggest anything that might make this situation worse. So how do we enhance circadian functioning in old age while protecting the eyes? Our best answer at this point is to increase light levels at appropriate times of the day, while using glare-reducing diffusion filters to keep the light comfortable and making sure that blue wavelengths are not overrepresented. For example, use soft-white lightbulbs rated at 2,700 or 3,000 kelvin. (For a more detailed description, take a look at Chapter 7.)

Outdoors and In

Some parts of the world, and some times of the year, offer more hours of bright sunlight than others. There is not much we can do about that, aside from deciding to move to Arizona or the French Riviera. Indoor room light, however, is an environmental factor under our direct control. Unfortunately, many older people use that control in exactly the wrong way. They adopt a lifestyle that minimizes their exposure to both indoor and outdoor light.

Many older people doesn't mean *most*, of course, and indeed a study of healthy people averaging age sixty-six found they received more daylight exposure than a younger cohort averaging age twenty-three. Most of these older folks were retired and healthy, with time and energy to spare, while almost all the younger ones had full work schedules that would keep them indoors. However, as we consider increased age with the likelihood of infirmities, medications, and difficulties getting around, the likelihood of lower light exposure multiplies.

But why do these older folks darken their daytimes? One reason is that they dislike wasting a penny of their limited and fixed income on something they see as a pointless extravagance like brighter light. If you can see where you're going, what more do you need? Also, lamps and ceiling fixtures that are not properly designed can create unpleasant glare that becomes still more bothersome as our eyes age. So we turn down the lights. Their children may repeatedly tell them to put in brighter lightbulbs and to get outside more, but it's a difficult sell. And the problem worsens after moving to an elder care facility that confines residents indoors for most of the day.

Our inner clock expects to receive contrasting signals of light and dark that indicate the sequence of daytime and nighttime. Those who spend the day in dimly lit interiors are depriving their clock of that essential cue. This in turn weakens the ability of the inner clock to regulate the daily cycles of attentional and metabolic processes. Something as simple as brightening the living room, making it function as an extension of daylight, can help maintain energy through the evening, forestall premature sleepiness, and increase the length of quality sleep during the night.

Marijke van P.'s Story

I am a single woman, I am seventy-six years old, and I obviously retired years ago. I used to work as a librarian and loved languages and reading, which inspired me to study Dutch literature. After I got my degree, I taught in higher education for many years.

And I have been depressed, very depressed.

My first episode was in 1994–1995. Two relatives were seriously ill and hospitalized. Another close family member passed away that year. In the months that followed, I was not doing so well. Tests showed that my thyroid was not working properly and I have been taking Synthroid ever since. My doctor also prescribed Paxil, and after a few months I started feeling much better. I continued to have regular checkups with my GP.

My second major depressive episode started in 2006. My youngest brother, a boy with Down syndrome, had died a few years before, after a long period of serious illness with gradual deterioration. We were very close and he viewed me as his mother.

Of course, I talked about all this with my primary care doctor and she felt I needed specialist treatment. For a while I saw a psychologist; I also took prescription medication from a psychiatrist. I felt ill. I was totally exhausted all day—I did not have the energy to do anything and just felt like sleeping. At night I felt a little better and started going to bed at increasingly later times, but in the morning I felt tired again. I just wanted to sleep. It took every ounce of my willpower to get out of bed; often I did not get up until noon. During the day I frequently had to fight an overwhelming urge to sleep, which slowly subsided.

At one point, the medication proved no longer effective and ECT therapy seemed to be the only chance of getting better. That is how I met Dr. Harm-Pieter Spaans, at the Parnassia Clinical Center for the Elderly in The Hague, the Netherlands. He suggested I try a new method first—bright light therapy. I took the opportunity with both hands, because I had been rather anxious about the ECT therapy. And I particularly liked the fact that I could be treated in the comfort of my own home.

I had my first treatment a few weeks later. Dr. Spaans had told me all about light therapy, how and why it worked, and a light box was then placed in my house. Since I live alone, Dr. Spaans had said I could call him night and day, if necessary. In my first week I had to stay awake one night and then sleep one night. Actually, this stage took more than two weeks. After a wake night, I sat in front of the bright light box for thirty minutes at 7:30 AM. I also took one Paxil after breakfast. At night I turned in at 9 PM. The next morning I sat in front of the light box again, also at 7:30 am, and took another Paxil. That was my new sleep/wake cycle.

The wake nights were sometimes not a problem and sometimes very hard, particularly after 4 AM. I usually did something that forced me to stand up and walk around in order to fight the cold, the urge to sleep, and the fatigue. I baked quite a few cakes in the middle of the night. And I did whatever I could think of—reading, sewing, cleaning out closets, doing jigsaw puzzles, and playing computer games, etc.

After sixteen days, after a sleep night, some kind of miracle happened: I felt like my old self again! I immediately had to switch to another cycle—getting up at 7:30 AM and going to bed at 11 PM. And I have been sticking to it ever since.

I am doing very well now. In the past years I have had two wake therapy weeks to readjust my sleep/wake cycle. I still have regular checkups, I still take one Paxil every morning, and I still use thirty minutes of light therapy in the morning. The only thing I find difficult at times is that I always have to get up and go to bed at the same time every day. But it is definitely worth it!

"Early to Bed . . ."

It is not hard to describe what a good sleeping pattern should be. You would go to bed at a reasonable hour, neither too early nor too late, and fall asleep easily. After sleeping through the night, with no interruptions, you would wake up alert and refreshed. And we know of two factors that, acting together, help bring this about. One is *sleep pressure*, which builds up during the time you are awake. The other is a switch in the internal clock signal from alertness to sleep onset. (For a more detailed explanation, take a look at Chapter 2.)

You might think that the closer we get to bedtime, the easier it would be to fall asleep. Not so. The inner clock continues to keep us in awake mode during the evening, even while sleep pressure is building up. These hours are sometimes called the "forbidden zone," because it is normally so difficult to fall asleep during them. Then, when the clock shifts into nighttime mode, we feel that wave of intense drowsiness that sends us to bed for the night. But if the circadian rhythm has weakened with age, sleep pressure can come to the fore earlier in the evening. This creates a vicious cycle of earlier bedtime, shorter sleep, and earlier awakening. Paul R., whose story opened this chapter, is an example of this.

Paradoxically, early bedtime in the elderly may also come about in spite of reduced sleep pressure. The culprit here is excessive daytime napping. With each nap, the sleep pressure dribbles away and has to build up again once the nap ends. The napping itself may reflect a mixture of pure boredom or inactivity, physical frailty or chronic illness, depression, and a weakened circadian signal for daytime alertness. That last factor, the weakened alertness signal, also eliminates the evening forbidden zone. The result? Even with all that napping and the reduced sleep pressure that results from it, the senior is likely to want to go to bed at an abnormally early—and unhealthy—hour.

Alzheimer's and the Inner Clock

When the inner clock is working as it should, it improves daytime cognitive functioning and mood, reduces daytime napping, and promotes uninterrupted sleep during the night. But inducing the clock to work properly in older people is a major challenge. The challenge is all the greater for elderly patients with neurodegenerative conditions, such as Alzheimer's disease, that affect the circadian clock, melatonin production, and sleep integrity itself. For Alzheimer's patients who are still at home, a major reason they end up hospitalized is that they stop sleeping soundly at night. This poses a grave risk of night-wandering and accidents, and arouses fear and desperation in family members.

In a landmark clinical trial in the Netherlands, patients with dementia spent most of their daytime hours under enhanced artificial lighting in hospital group rooms. Then in the late evening they were given melatonin. The results were remarkable. Their nighttime sleep coalesced, and they showed more consistent daytime activity, more positive mood, and a slowing of cognitive decline.

Simulation of gradual dawn and dusk twilights in the bedroom also helps. In a Swiss study, Alzheimer's patients went to sleep in the presence of a dusk signal and woke up during dawn signals. Within three weeks they were sleeping an hour longer during the night, and their daytime restlessness improved as well.

Gabriel R., seventy-eight, was living in an elder care facility. Unlike Alzheimer's patients, he was able to follow conversation despite short-term memory loss, but his sense of reality had slipped, and he was wrenched by long periods of deep depression. He generally fell asleep at 9 PM but woke up at 11 PM and binged on carbohydrate-rich foods. Then he would fall into a fitful sleep around 1 AM and wake up for the day three hours later, feeling awful. We placed a dusk simulator at Gabriel's bedside. It stayed bright from 9 to 11 PM to discourage him from going to sleep early. Then it gradually faded to darkness as he fell asleep. He began waking up two hours later than before, at 6 AM. For the first time he agreed to go to the dining room for breakfast and to take a morning walk. He also found a new enjoyment in daytime activities. So even something as simple as enhanced lighting during the evening corrected a sleep onset problem that is all too familiar to many elderly, whether or not dementia has set in.

Shedding Light on Parkinson's

The devastating effects of Parkinson's disease on motor control, sleep, mood, and thinking have made it one of the most feared maladies of the elderly. It does not always wait until the later years to appear, as the example of actor Michael J. Fox shows, but most cases occur in those over fifty. The motor symptoms of the disease, such as tremor, slow movements, and muscular rigidity, are the most obvious, but it can also lead to difficulties with planning, cognitive flexibility, abstract thinking, and impulse control. Sleep problems linked to Parkinson's include daytime drowsiness and insomnia. No cure is known. The standard treatment is long-term medication with drugs that help replenish the neurotransmitter dopamine in the brain. These medications generally slow down the worsening of symptoms, but results are quite variable from patient to patient.

There are strong hints that circadian rhythms have a connection to Parkinson's, but there has not been nearly enough research to verify this possibility. One study looked at Parkinson's patients who were taking the usual dopamine-based drugs for their condition. The researchers found that the patients' inner clocks were shifting the timing of melatonin release by the pineal gland. Exactly what this finding means, however, is still unclear.

Dr. Gregory Willis, in Melbourne, Australia, has taken a big step further. He is using bright light therapy, along with medications, in a large group of Parkinson's patients who have been followed for up to eight years. As we might expect, mood and sleep improved rapidly, within weeks after starting light therapy. More strikingly, the patients' motor disturbances also improved significantly, although more gradually, over months and years. Many of the patients were able to reduce their medication levels without their symptoms worsening. Meanwhile, patients who did not receive light therapy had no significant reduction in insomnia or depression and showed worsening motor symptoms over time. One important finding is that those patients who consistently used light therapy every day showed the best improvement by far, while those who used it only off and on showed a patchy trend.

This work is very new. The effectiveness of light therapy in treating Parkinson's will have to be tested further in independent clinical trials. If the positive results hold up, researchers will need to chart the best ways to use light treatment. How bright should it be? For how long at a time? At what point in the day relative to the patient's chronotype? Are there any negative side effects that should be taken into account? It is still too soon to say that all patients with Parkinson's should begin using light therapy—and certainly not on their own. But at the very least, the medical community and all those who are concerned about Parkinson's should be aware that they may soon have a major new tool for treating the ravages of this much feared disease.

Melatonin and Aging

The circadian clock is not the only component of the sleep process that changes as we grow older. With age, the pineal gland gradually produces lower nighttime levels of melatonin. The result is a diminished physiological signal of darkness. As a result, it is harder for the body to switch from day-appropriate to night-appropriate metabolism, blood pressure, hormone secretion, and so forth. This may be one reason sleep becomes lighter and more easily interrupted with age. Not everyone shows this decline, however, and we do not yet understand who is fated to have their nighttime melatonin levels go down in this way. Lower levels of melatonin during the night do not necessarily lead to poor sleep, but in many cases replenishment with appropriately timed doses of synthetic melatonin has shown remarkable benefits.

Lili S. was a seventy-two-year-old widow who loved her work as a lab technician, and decided never to retire. On weekends her joy was in gardening around the house where her family had lived for more than fifty years. She was just the opposite of someone suffering from depression, but she did have a problem. She was up and down all night. Her light, broken-up sleep left her with uncharacteristically low energy during the day. This kind of sleep pattern can strike people at any age and for a wide range of reasons. In this case, our guess was that her inner clock was putting out too weak a signal to keep her in sleep mode. Could that be reinforced by a microdose of controlled-release melatonin? We suggested she take a single tablet daily at 9:30 PM, her regular bedtime. Our hope was that the melatonin from the tablet would give just enough of a boost to her own melatonin already circulating from the pineal gland. (For a fuller explanation about microdose melatonin, take a look at Chapter 8.) Astonishingly, it worked in one day! Her sleep immediately coalesced into a single nightlong episode, and her smile the next day at work was wider than her coworkers could remember.

What About Naps?

One question about seniors and sleeping that has generated a lot of discussion is whether to take an afternoon nap. Some researchers say that siestas improve alertness and cognitive functioning without affecting nighttime sleep. Others insist that napping in the afternoon carries a risk of insomnia, for both older and younger people. Most would agree that those who decide to take an afternoon nap should make sure it is:

- *early*—ideally, before 3 PM
- *short*—no more than half an hour, and possibly as short as five to ten minutes
- *restful*—in a quiet, dark, and comfortable setting

Those who nap much more than half an hour are likely to suffer from grogginess and lack of alertness when they wake up, and those who nap much later than 3 PM are more likely to have problems getting to sleep that night.

Stephen W., seventy-five, who was depressed, generally napped for an hour after dinner. He woke up fully alert but then found it impossible to get to sleep again before 3 AM. Even when he tried, he couldn't manage to control his evening napping. We soon found out why: It was triggered by cocktails before dinner and wine during the meal. This was an indulgence he had enjoyed since his early twenties. Stephen was also an owl chronotype. He had had trouble sleeping normal hours since childhood. When he agreed to reduce his evening alcohol intake, and even skip it on some days, the evening nap vanished and he was able to get to sleep an hour earlier.

"Earlier"—2 AM—was still very late by his wife's standards. It was clear that we had to reset Stephen's inner clock. Our aim was to move his nighttime wakefulness zone earlier, from after midnight to during the late evening. He began a dual course of chronotherapy, with a microdose of controlled-release melatonin at dinnertime and bright light therapy when he woke up in the morning. At first, we set the light therapy for 11 AM, but gradually he managed to shift it to 9 AM. Between eliminating his evening naps and resetting his inner clock, Stephen was soon able to begin sleeping from midnight to 9 AM, with only occasional nighttime interruptions. And not long after he began bright light therapy, his depression mood ratings rose from "very bad" to "near normal."

Steps to Take

If any of these descriptions of common sleep problems seem to fit your experience as an older person, or that of an older person you are concerned about, take heart. There are effective steps you can take to get back on the road to healthful, refreshing sleep. Some of these are basic measures of sleep hygiene that most of us know but too often ignore.

DON'T

- Drink caffeinated beverages (coffee, tea, cola, cocoa) in the second half of the day.
- Have dinner within three hours of bedtime.

- Exercise during the evening.
- Smoke near bedtime.
- Drink alcoholic beverages after dinnertime.
- Read from a backlit e-reader or similar screen devices within two hours of bedtime, unless the blue output can be controlled.
- Read, snack, or watch TV while in bed.
- Get into bed if you do not feel sleepy.

BUT DO

- Try to keep to a regular bedtime, even on weekends.
- Develop a relaxing pre-bed ritual, such as a warm bath or quiet music.
- Keep your bedroom as quiet and dark as you can.
- Attend to medical conditions, such as apnea, restless legs, and urinary problems, which interrupt your sleep.

For many older people, simply putting these basic steps into practice is enough to bring about a good night's sleep. If not, it is time to consult your physician or a sleep medicine specialist. Persistent insomnia should *not* be accepted as a normal part of aging.

Think carefully and critically about the use of sleeping pills to treat insomnia. This is especially relevant for seniors, who buy almost half of the prescription sleeping pills sold, even though they make up less than one-fifth of the population. There is a place for these powerful medications, but in our view they are best used only after other therapies have failed to help, and even then only as a temporary measure. Take the case of Stephen W., whose sleep improved as he introduced microdose melatonin and light therapy, and reduced his evening alcohol intake. As this happened, he was able to achieve a long-standing personal goal: tapering the dose and finally quitting nightly Ambien.

Sleep disturbance is not an inevitable consequence of aging, but it often seems secondary when compared with acute and chronic illnesses. True, some illnesses can cause insomnia (or oversleeping). But the opposite can also hold. Intelligent sleep management can help lighten other medical and psychological burdens and reduce the need for taking multiple prescription drugs—all steps toward increased quality of life.

Getting to Better Sleep

The medical community as a whole is just beginning to be aware of the ways the inner clock is connected to sleep difficulties. Indeed, the first treatment manual for clinicians was published only in 2009 (see *For Further Reading*, p. 303), and with all the new concepts and procedures it takes some serious study. Doctors will need to monitor their chronotherapy patients closely and critically, especially since dosing and timing of light and melatonin may need frequent adjustment at the start. Doctors will have most success if patients clearly describe the issues for which they think chronotherapy would help.

The first step in putting chronotherapy to work is to decide what the most important sleep difficulty is. Is it:

- Falling asleep well before normal bedtime and waking up before morning?
- Being unable to fall asleep until well after normal bedtime?
- Sleeping too lightly, with frequent wake-ups?

Each of these has a different relationship to the inner clock and requires a different treatment approach. Here are some suggestions.

Falling Asleep Too Early

- Go outdoors for a walk at midday rather than early in the morning.
- Have your major meal of the day at lunchtime.
- Cut down on napping, especially in the afternoon and evening; move around, do stretches, turn up the lights.
- Move dinner earlier.
- Turn up the room lights during the evening, or use a bright light therapy box about an hour before you usually get sleepy, for fifteen to thirty minutes.

Falling Asleep Too Late

- Keep your bedroom dark until you wake up. (Early morning light through the windows can have the paradoxical effect of maintaining the late-sleep pattern.)
- Consider installing a dawn simulator in your bedroom, and set it to rise to maximum light level at your natural wake-up time, even if that is in the late morning.
- Consider using a bright light box shortly after your final awakening. Whether you use a dawn simulator, light box, or both, begin shifting the lighting schedule and wake-up time earlier, in fifteen-minute steps, as soon as you feel comfortable waking up at each step.
- If you find the wake-up light approach difficult, add microdose controlled-release melatonin six hours before bedtime, and use a reminder cell phone or wristwatch alert to stay on

schedule. Move the melatonin earlier in parallel with morning light schedule.

- Finish dinner at least three hours before bedtime.
- Try to minimize napping, especially in the late evening: move around, do stretches, avoid all alcohol after dinner.
- Dim your evening room light, using dimmer switches or plug-in dimmers that rest on the floor, but keep the light level high enough to be able to see clearly throughout the room.
- If you are sitting up close to the TV in the evening, wear blue-blocker glasses to reduce the activating effect of screen light.
- If you are spending time at the computer in the evening, install the f.lux application (see *Resources for Follow-up*, p. 301) to automatically reduce blue emission.
- If your feet are usually cold when you get into bed, take a warm shower first, or wear socks.
- If you are following our chronotherapy procedures and also taking sleeping pills to get to sleep, watch for feelings of sleepiness *before* you take the sleeping pill. This is a signal that the chronotherapy is superseding the effect of the sleeping pill. It is time to talk with your doctor about reducing the dose or eliminating the sleeping pill entirely.

Sleeping Fitfully

- Eliminate alcohol after dinner.
- Take a microdose of controlled-release melatonin (see *Resources for Follow-up*, p. 301) twenty minutes before you go to bed.

- Keep your bedroom dark.
- Bladder control: If you are waking up at night to urinate, install amber-colored night-lights in the bathroom and hallway (see *Resources for Follow-up*, p. 301) instead of switching on regular bright lights.
- If sleep improves, and you have been using sleeping pills, it is time to talk with your doctor about reducing the dose or eliminating the sleeping pill entirely.
- If sleep improves but you are groggy in the morning, try one or both of these steps: (1) take only half of the melatonin tablet; (2) take the melatonin tablet two hours before bedtime, instead of twenty minutes.
- If you find you stay sleepy all day while taking the melatonin, stop using it. Melatonin is not going to be the solution to your problem.

Sleep disturbance is not an inevitable consequence of aging, but it often seems secondary when compared with acute and chronic illnesses. True, some illnesses can cause insomnia (or oversleeping). But the opposite can also hold. Intelligent sleep management can help lighten other medical and psychological burdens and reduce the need for taking multiple prescription drugs—all steps toward increased quality of life.

PART 5

Chronotherapy in Your Life

15. Coping with Shift Work

Just a few generations ago—and even today in many industries and many parts of the world—most people had workdays that started around dawn and finished around dusk. This was the case for farmers and factory workers alike. The most important reason was straightforward: Once night came, it was too dark for most kinds of labor. In Dickens's *A Christmas Carol*, at the end of the day Scrooge was reluctant to spend the money for a candle, even if it would have let him squeeze a little more work out of his clerk, Bob Cratchit.

There were some exceptions, of course. Trains and ships that kept moving through the night needed railwaymen and sailors to operate them. It was always pitch-black down the mine, so the time of day didn't much matter. Shuttered warehouses, shops, and homes called for night watchmen to guard them while other honest folk were asleep. But for most, sunup and sundown defined the natural limits of the working day.

Then came Thomas Edison. Once artificial electric light became plentiful and cheap, the rules began to change. What sense did it make to have expensive machinery sitting idle for two-thirds of every day? If there were customers on the streets late in the evening, why not keep your store open for them? And of course the gas plants and electric plants *had* to operate around the clock, to feed the unending need for energy. Gradually, urged along by wartime booms and the spread of globalization, a 24/7 world economy came into being. With it came the need to have people available to work at every hour of the day and night. A system quickly developed to meet this need. It is called *shift work*, and it has had terrible consequences for millions of people.

The underlying problem with shift work is that it comes into direct conflict with the normal human circadian rhythm, as set by our inner clock. Whatever our chronotype, whether we are larks, owls, or hummingbirds, our ability to think, act, and pay close attention to what we are doing goes through daily cycles of highs and lows. If our work schedule calls for maximum alertness, but we are at a point in the day or night when we can barely keep our eyes open, disaster lurks. A moment's inattention may cost you a finger, or cause a multi-car pileup, or threaten thousands or millions of people.

The nuclear accidents at Three Mile Island and Chernobyl, like the catastrophic leak of poisonous gas at Bhopal, India, happened late at night. So did the grounding of the supertanker *Exxon Valdez* in the coastal waters of Alaska, which created one of the worst environmental disasters in U.S. history. Hundreds of thousands of seabirds, seals, sea otters, and salmon were killed by this massive oil spill. Airline pilots sometimes fall asleep at the controls, and so do commuters on their drive to or from work. Chronotherapy does not offer any magic solution to these hazards, but by understanding what they are and why they occur, we can find ways to reduce the risk they pose.

Shifting Times

It sometimes seems that there are as many different shift work schemes as there are organizations that use them. Five-day weeks, seven-day weeks, or even eight-day weeks? Eight-hour workdays, nine-hour workdays, or longer still? Do you stay on the same shift for long periods of time, or do you constantly rotate through different shifts? Do the different shifts overlap or not? There is even a system, used on U.S. Navy ships at sea, that involves "days" that last eighteen hours instead of twenty-four!

The most familiar pattern, and the one that we will give the most attention to, divides the twenty-four-hour day into three shifts of eight hours each. Employees generally work five days a week, with days off ("weekends") that may change regularly to provide 24/7 coverage. Workers usually stay with the same shift for a lengthy period of time instead of cycling through them. Typical examples of the three shifts are:

- First shift, or morning shift, from 6 AM to 2 PM
- Second shift, or swing shift, from 2 PM to 10 PM
- Third shift, or graveyard shift, from 10 PM to 6 AM

Each of these creates a different set of problems for anyone with a normal sleep/wake pattern.

- For those on the first shift, getting to work by 6 AM means getting up very early, at 4 or 4:30 AM. That in turn makes sleep pressure build to a high level by early evening, long before the inner clock gives the sleep onset signal. You may have to be asleep by 8 or 8:30 PM to get enough sleep, but that does not mean it will be easy to do.

- The second shift fits better into most people's daily cycle. That is surely why workers see it as the most desirable. However, for those with families, it means no time with the kids, who are at school during the free morning hours and in bed by the time the swing shift worker gets home.
- Working the third shift is hardest on practically everyone. The work hours fall solidly within the normal nighttime sleep period, so the urge to sleep grows stronger and stronger as the hours creep by. Working this shift also forces the employee to catch up on sleep during the day. This is not an easy matter in the best of cases. It is no wonder this shift has the highest incidence of illness, work errors, and job accidents.

It is almost as if the shift system was designed to take advantage of people who already have sleep disorders. Extreme larks, those with *advanced sleep phase disorder* (ASPD), can probably handle the first shift fine. After all, even when they are not working, they generally wake up very early, 4 AM or so, and need to go to sleep at a correspondingly early hour. As for the graveyard shift, what could be a better fit for an extreme owl, someone with *delayed sleep phase disorder* (DSPD)? The wee hours of the night are when they normally function best. What this leaves out is the stress their shift work—*and* their sleep disorder—imposes on their family and social life.

We suspect that many of those who are attracted to night work already have circadian rhythms that would put them solidly in the owl category. Consider Cliff M., twenty-eight, whom we met in Chapter 8. He could not get to sleep until about 7 AM. It had been this way since he was a teenager. He was a healthy and successful young scientist—except that he didn't arrive at the lab until 5 PM, just as his coworkers were preparing to leave. This made collaboration difficult or impossible. But if he had been a musician whose band regularly went on at 11 PM or later, this would not have been a problem at all.

José's Story

José P. is fifty and has been a doorman for twenty-five years. He has always been on the night shift, from 11 PM to 7 AM. Tenants in his Chicago apartment building say he is consistently pleasant and cheerful.

José likes being on the night shift. "I'm off work at seven [AM], home by eight, and then I'm up until two or three [PM]. Then I go to sleep for four or five hours, then ready for work again. Sometimes it's hard and I don't want to get up, but I'm so used to it that I just get up! And once I'm up, I'm wide-awake—I'm wired up. The only time I'm moody is when my stepson and his friends are playing games and making a lot of noise in the hours when I'm trying to sleep.

"When I'm leaving the shift, it's still dark," he adds. "But I feel refreshed when the train comes out after the tunnel, especially during the winter hours. I fall asleep in the tunnel and have a half-an-hour nap. But when we come out into the sunlight, I'm wide-awake. I come home and I don't want to go to sleep, I play handball or something. When I'm playing, I use a lot of energy, and when I go home I fall asleep right away. It's harder to sleep in the summertime, because I don't want to miss the day. I love to go to the beach and play handball, bike ride. I'm very active. I get home and go to bed about two or three in the afternoon and I'm up at seven or eight.

"I have a second job from two to ten on Saturday and Sunday," he continues. "So I do thirty-two hours in two days. On Monday, I'm so tired that even though it's one of my days off I'm so exhausted, my body is so burned out that I'll sleep till seven or eight in the evening. Usually I just get four hours' sleep a day. I took the second job for holiday money almost four years ago—but they're so nice that they offer me money when I say I'm going to quit. They treat me very well, so I just stayed.

"My body is used to the shift. I lie down sometimes on the bench to relax my body so my back doesn't hurt, but I'm wide-awake." He doesn't feel he needs anything to help him stay up. "Just coffee, usually when I'm working I have five cups a day. That's when I'm doubling, and doing the Saturday/Sunday extra shift. When I'm not doubling, I only drink about four. And I never have coffee two hours before I leave the shift at seven AM, or I won't be able to sleep in the afternoon. I'll just toss and turn."

Working the night shift is not *all* positive for him. "Where it bothers me is that every weekend—Friday, Saturday, and Sunday—I'm on, so I can't do anything. My friends don't even call me anymore—they say, 'Don't even ask José, he can't do it.' So I miss birthday parties, weddings, all the family get-togethers. I ask people to switch with me, and they won't. Several times I have asked [my boss] to change my shift, because of something special I needed to go to, and he wouldn't. I've been working a long time, and he should be fair—but he won't do it. He won't budge."

José has some thoughts about the future. "I'm not retiring yet. I'm going to continue until I retire. [Then] I'll be able to go to all those things that I have to miss out on. Anybody wanna go party? I know I won't be tired at night! I'm just a night person—I can't function during the day [by the afternoon], dealing with everything that's going on. At night, I'm at peace."

Human resources managers keep trying to design shift systems that maximize the twin factors of alertness and productivity. Some are quite ingenious. Others are so convoluted that even after months in place, employees have to consult a chart to figure out when they are coming to work next. And all these systems run up against a basic reality. To function at peak efficiency, our biological systems require not just enough sleep, but enough sleep during the right part of our circadian cycle. When we don't get it, the results include less energy, lowered alertness, weakened cognitive ability, and for many, chronic and serious depressed mood.

Problems and Personal Costs

In industrialized countries, as many as one in five workers works either at night or on rotating shifts. These workers have jobs in medicine, transportation, manufacturing, and public safety. Then there are those employed by fast-food restaurants, all-night groceries, and other enterprises that provide needed services to night-shift workers. In surveys, up to two-thirds of shift workers complain of excessive sleepiness and report that they fall asleep on the job at least once a week. Other problems linked to shift work include:

- Insomnia
- Trouble concentrating
- Ongoing fatigue
- Chronic indigestion
- Increased irritability
- Depressed mood
- Falling asleep while driving
- Difficulties with personal and family relationships
- Higher absenteeism
- More workplace accidents

Evidence is accumulating that night-shift work can have major consequences for physical health as well. One recent Swedish study found significantly higher rates of multiple sclerosis in night-shift workers who had worked at least three years by age twenty. A Finnish study of former airline shift workers identified a higher rate of the metabolic syndrome that is linked to coronary artery disease, stroke, and type-2 diabetes. (This association turned up only in the male employees, not the females.)

Other studies have found increased cancer rates in night-shift workers. As a result, the World Health Organization has provisionally identified the circadian rhythm disruption of night-shift work as carcinogenic. Some researchers suggest that it is specifically the exposure to nighttime light that is to blame for the higher cancer risk. It is well known that light exposure during the sleep cycle suppresses the pineal gland's production of melatonin. If we assume that melatonin helps protect cells from damage by carcinogenic free radicals, then the suppressed melatonin levels might trigger cancer growth.

But when we consider sleep loss, forced activity during the circadian night, flipping back to normal schedules on days off, irregular diet and mealtimes, and the ongoing effort to stay awake at work and then at the wheel going home, we have more than enough system stressors to account for the health damage.

Anthony's Story

Anthony M., forty-eight, like José P., happens to work as a doorman, not in Chicago but in downtown Boston. He lives with his wife and two children, a son, eighteen, and a daughter, thirteen. He works on a flex shift. Thursdays, Fridays, and Saturdays he is on the afternoon shift, from 3 PM to 11 PM. Sundays and Mondays he works the night shift, from 11 PM to 7 AM.

"With the day-to-day changes in the time, it's very difficult," Anthony admits. "Monday and Tuesday I'm forced to go to sleep earlier. The kids say, 'You must be working tonight, because you're grumpy.' I sleep from three PM to seven PM. Wednesday, I'm too wired to sleep and I go to bed about nine PM and sleep till six AM. I take the children to school and do chores until about one PM, then I get here at two PM so I can have coffee and not rush when I start work at three PM.

"It beats me down, this schedule. On Saturday, I'm working from three to eleven. Sunday I'm supposed to be off, but I'm already thinking twenty-four hours ahead. I've got to plan my day. I'm in a bad mood and grumpy, because before I come here at eleven Monday evening I'm trying to sleep. I don't usually get to sleep before three, then I have to be up at seven to have dinner and shower and have time with the kids. And at nine PM I have to leave to come to work. I'm going to bed only for a few hours."

This is not Anthony's first experience with shift work. In his twenties, he was a manager at a fast food restaurant where his schedule changed every week. The first week, he worked from 11 PM to 7 AM, the second week from 3 PM to 11 PM, and the third week from 7 AM to 3 PM. Then the pattern repeated. "I didn't like it, but we were young," he says. "Two out of the three schedules were okay. I could get through the week and look forward to the following week when the schedule changed. It wasn't as hard to adjust as it is now."

When Anthony has to work the night shift, "I leave home at nine PM. I'm dragging. And the overnight shift is *boring*. The more boring it is, the more I tend to want to doze off. It's too dim. There's very little traffic, it's cozy and warm, the light is soothing. It gets you in the mood to go to sleep. You know, I want to keep busy, and we have chores. So I clean the floors, and that keeps me sharp for a while. And I drink coffee. I'm not comfortable until I start cleaning. At least it's a task. It takes about four hours to do it, but it helps keep me awake.

"My body is so used to a day schedule that the night shift is really hard," Anthony says. One reason is that he is naturally a lark. "I was always up before my mother would get up. She never had to wake me for school. I'd go to bed around nine-thirty and get up at six. I've been doing it since I was fifteen years old. On vacations I enjoy the time. I'm on my regular schedule, going to sleep at ten PM and I wake up at six AM. My body just wakes up—I don't need an alarm clock, or a rooster."

When his shift ends at 7 AM, "I've got my second wind and I'm ready to roll. I stay up as if I'm going through a regular day. I do my chores, the Internet, mail. It feels right to me. It fits my body schedule. When I do go to sleep in the afternoon, the daylight affects me. I've got blinds, but it is difficult to catch sleep."

Anthony resents the way his work schedule keeps him from advancing in his occupation. "In my union, they give training courses in carpentry, electrical work, heating. I'd really like to take some of those—I like to keep learning. But they're scheduled from six to nine, or seven to ten, and my schedule prevents me from taking them. To me, these courses would be a real benefit."

The worst part is the way shift work interferes with Anthony's family and social life. "The schedule puts on a lot of stress, because the kids and my wife would like to have me around more. When it comes to family events, or parties or anything with friends, I'm not available. I tell my wife and my kids to go have fun, but she'd rather I could go, too. Over the summer, I'm not available for picnics, barbecues, and the beach. I never get Memorial Day, or Labor Day, or Thanksgiving off."

Anthony is resigned to staying on the flex shift schedule "until there's an opening for another shift or I work somewhere else. It's an unwritten rule. There's seniority, so you just have to wait for everyone else to reject the shift that you might want, and then you might be able to change. I read this article that the president lives two years for every year [he is in office]. I think for a doorman, it adds a month every six months. What I'm really saying is, it reduces your life span."

Trying to Cope

Those who have spent any length of time doing shift work usually develop some scheme to help them deal with the fatigue and unreality of their work life. They may, for example, set aside time for a nap before going in to work or right after coming home, and try to fit their main block of sleep into some part of the day that seems to make more sense. What generally doesn't have a place in their calculations is the role of their biological rhythms.

Even if you have had a nap or your main sleep before going onto the night shift, your circadian clock would still issue its sleep-onset signal and the pineal gland would still deposit melatonin into the bloodstream. Sleeping beforehand reduces sleep pressure, which may make the first few hours on the job easier to handle, but you will still have to fight to keep your eyes open during the wee hours. Lots of shift workers depend on coffee, sugary snacks, or pills to help them survive the night, but their metabolic aftereffects can linger in the system and make it harder to get needed sleep once the shift is over.

After a night's work, there is the disorienting experience of going home as the sun is rising. Some shift workers try to sleep as soon as they get home, but this is far from easy. Your inner clock has just issued its wake-up signal, putting you at the start of your circadian day. You may try to take advantage of the pent-up sleep pressure from your night awake, but the sleep you get is going to be shorter than normal and unsatisfying. So you sleep briefly, out of exhaustion, and put off your main sleep until later in the day. These brief periods of sleep—naps, really—have the downside of temporarily lowering sleep pressure. As a result, the main sleep later in the day doesn't last as long as you would like, and you gradually build up a sleep deficit.

Then come the days off. These are supposed to give you a chance to recover. Unless you have no social life at all, you will probably want to snap back into a normal sleep/wake cycle. That way, you have at least a chance at getting some quality time with family and friends. One shift worker told us, "It makes me furious when the boss or customers phone me on weekends, even if it is important. I need that solid time with my wife and kids just to stay sane." And for many shift workers, even time on days off is not so solid, because they are sleeping longer than normal to make up for the sleep deficit they accumulated during the workweek.

Shift workers with children have an additional layer of problems to deal with. More and more childcare facilities offer extended hours or even around-the-clock services for parents on shift work. But these are expensive, even unaffordable, and many parents have qualms about dropping their kid off somewhere in the middle of the night. Managing childcare becomes especially complicated when both parents are on shift work. Some couples arrange their shifts so that the one on the way to work can hand the children over to the one coming off work. It is easy to imagine the many ways things will go wrong with arrangements like this. And even when everything goes as planned, each of the alternating caregivers will need to sleep, no matter how much they may want to provide good supervision and get some quality time with the kids.

What Can Chronotherapy Do?

"If the reason I'm having so much trouble handling the night shift is my inner clock, why don't I simply reset it and turn my personal nighttime into daytime?"

What this suggestion amounts to is deliberately giving someone a severe case of delayed sleep phase disorder. There may be situations that could justify a radical step like this. Facilities such as oil refineries and nuclear plants operate continuously and call for constant supervision. Forest fires, water main breaks, and medical emergencies show no respect for nine-to-five conventions. Sneak attacks generally occur during the hours just before dawn. Even customer call centers typically have to deal with clients many time zones away. There may well be people, extreme owls, who would be drawn to the prospect of living their lives so radically out of sync with those around them, but we suspect there are not that many.

The inner clock cannot flip quickly or frequently between nocturnal and diurnal patterns. Most night workers want to keep at least one foot in the daytime world. They have families that live on a daytime schedule, kids who need parenting, friends they would like to see at least occasionally. This makes them committed to shifting into daytime activity on days off, whether it is every few days (for flex shift workers who put in twelve hours on successive nights each week), or weekends, or on week-long "vacations" after several weeks of consistent night work. They know they will suffer during the week, but they will also have the satisfaction of living more like normal humans on their days off.

There have been some attempts to use techniques of chronotherapy to help night-shift workers. In one of the first, which we carried out at Columbia, night-shift news writers at a national TV network did light therapy at home just before leaving for work. The effect of this was to shift their internal clocks later, in the direction of delayed sleep phase, but not to create a complete day/night reversal. The light treatment did make most of the participants more alert when they started work. The excruciating pressure to sleep, which tends to come around 3 AM to 5 AM, was pushed later into the morning, which meant they found it easier to sleep when they got home from work. However, the timing of the light treatment was dictated by their work schedule, not by their individual chronotypes. Some of them had remarkable results, some only a partial result, and some none at all. The method did not catch on.

Other more ambitious research has tried to identify ways to use the circadian cycle to improve daytime restorative sleep without creating a twelve-hour flip-flop. Much of this work has been done in the laboratory. Participants come in for overnight sessions in which the sleep and lighting pattern is strictly controlled while their melatonin levels are measured. One insight from this research is that some participants do best by shifting their rhythms earlier, while others shift later. It makes sense that larks would find it easier to shift earlier, while owls would find the opposite easier. Now what we need is field trials, outside the laboratory, that look at real shift workers who need to deal with everyday obligations.

Shifting people's clocks later, as was done with the night-shift news workers, raises a problem: the need to avoid morning sunlight at the end of the shift. The burst of light can be subjectively excruciating, but worse, it falls at a point on the circadian cycle when the inner clock responds by resetting *earlier.* What is the use of resetting the clock later at the start of the night, and then resetting it earlier a few hours later? One possible solution to this problem is to wear blue-blocking glasses for the trip home. These maintain sharp visibility while filtering out the light that is most likely to reset the clock in the wrong direction.

Brighter Nights

The suggestion is often made that night-shift workers would cope better if their task lighting during the night shift were much brighter. Brighter light would raise their energy while working. It would also nudge the inner clock later, bringing it more into line with their activity schedule. But this nudging later would occur only during the first part of the night. If bright workplace light continued to the end of the night shift, it would instead nudge the clock earlier, the same way that dawn light does on the way home. And the time during the night when the effect reverses from nudging later to nudging earlier depends on each individual worker's chronotype. (For an explanation of chronotypes, take a look at Chapter 3.)

What about keeping the workplace bright during the first part of the night, then gradually reducing the light level? In principle this could work, but the timing would be very tricky. In effect, the lighting would have to be adjusted to the inner clock of the "average" employee. Those whose normal cycles were earlier or later than average would be affected differently, and not necessarily in a positive way. Another problem with enhanced whole-room lighting installations is that when they have been tested in field studies, the workers didn't like them and sometimes voiced their objections very loudly. It may be that individually controllable local-area lighting, for example, to spatially separate workstations would be better received.

Any pattern of nighttime lighting will have a powerful effect on the inner clock. As we have seen in earlier chapters, the biological mechanisms that keep our system synchronized with the twenty-four-hour solar cycle ordinarily depend primarily on the light we are exposed to at the beginning and end of our circadian night. Fading light at dusk, bright light at sunrise—that is all the system really needs or expects. When light exposure comes during the circadian night, when it is usually—or should be—dark, it activates the timing system and resets the clock. This is the rationale of light therapy for sleep disorders. Light therapy for someone with advanced sleep phase disorder—an extreme lark—will emphasize evening light exposure. Someone with delayed sleep phase disorder—an extreme owl—will need exposure at the *end* of the circadian night to prevent drifting later and to promote resets to a more normal sleep/wake cycle.

Managing Shift Work

There are so many versions of shift work, and so many ways one person's situation may differ from someone else's, that any suggestions we make are bound to be very general. There is one point, however, that applies to everyone, and that bears repeating. The most basic problem with shift work is that it does not fit well with the sleep/wake cycle that is such an important part of our biological functioning. Anything we can do to improve that fit will help us manage shift work better. Here are some ways we might do that:

- *Know your chronotype.* Are you a lark, an owl, or a hummingbird? How much so? If you have not already done so, go to the online chronotype questionnaire (see *Resources for Follow-up*, p. 301).
- *If you can choose, work a shift that works for you.* If you are a lark, you will probably be fairly comfortable on the early shift. Owls may find the night shift a tolerable fit. If you are in between, try for the day shift.
- *Get enough sleep.* Most shift workers get less sleep than they need. Practically everybody needs at least six hours, and most people need more. If you don't feel refreshed when you get up, you are not getting enough sleep.
- *Time your sleep to match your cycle.* No matter how tired you are, you will find it very hard to fall asleep at a time of day when your inner clock is giving wake-up signals. If your work schedule forces you to sleep during the day, try waiting until early afternoon, siesta time, when most people find it easier to doze off.
- *Develop a healthier lifestyle.* Shift work stresses your body, whether you rotate or stay with one shift. Exercise regularly before going to work (but not during the three hours before you intend to sleep). Eat a healthy diet, avoiding fatty and sugary foods. Use caffeine sparingly, and not during the second half of your shift. It washes out of the system slowly and may keep you from getting the sleep you need.
- *Protect your sleep.* A regular schedule and bedtime ritual can help cue your body that it is time to go into sleep mode. Block out noise, and likely sources of noise. Wear earplugs,

close the door, and turn off your cell phone. Ask family members and close neighbors to avoid vacuuming, mowing the lawn, playing video games, and other loud activities during your sleep time. Cover the windows or wear a sleep mask to keep out the daylight.

- *Use caution with alcohol and drugs.* Many shift workers relax with a drink or two, and many use medications to help them get to sleep or stay awake. All these practices call for caution. Alcohol can help you feel sleepy, but it also makes you sleep more lightly and wake up more often. Sleeping pills may help when used occasionally, but they should not be used long term (we discuss the reasons for this in Chapter 8). As for amphetamines and other "uppers" taken to keep sleep at bay, these are very powerful system stimulants that pose a real danger of addiction.

Some people find it easy to adapt to shift work. For others, it is hard, but they manage to overcome the difficulties by staying fit, developing coping skills, and enlisting the help of understanding family members. There are also those who never adjust to the physical, mental, and social upsets of shift work. For them, as for the night worker suffering from depression, the best advice—not always practical—is to avoid shift work altogether. Unfortunately, it is not likely that employers will listen favorably to such a request, even if it comes accompanied by a doctor's note. And for an employee to even *admit* such vulnerability may endanger job security and promotion. The suffering multiplies.

16.

Racing the Clock, Racing the Sun

If you are a fan of old movies, you have probably watched scenes that take place on a steamship. It may have been one of the renowned transatlantic liners—the *Normandie*, the *Queen Mary*, the *United States* (or, even more likely, the *Titanic*)—or maybe a much smaller, much grubbier tramp steamer. The film may have focused on the masses huddled in steerage, the elegant champagne-and-caviar set in first class, the square-jawed officers on the bridge, or even the grimy stokers down in the engine room. What the movie probably did not pay any attention to was the peculiar way the ship's clocks behaved. If the ship was headed west, at some point during the night the clocks jumped forward an hour. On eastward journeys, they jumped backward instead. The shifts were too small for the passengers to really notice, but the benefits were huge. By the time they landed in New York or Le Havre or wherever they were bound, their inner clocks were already well on the way to being in sync with local time.

Long-distance travel is not like that today. Even if the great liners were still in service, how many Americans have the leisure or means to spend a week to get to Europe or Hawaii and even longer to reach Asia or Australia? The world seems so much smaller today because modern air travel allows us to reach distant parts so much faster. But there is often a heavy price to pay—the familiar and dreaded experience of *jet lag.*

At a conference in Australia, we approached a large group of doctors who had come from around the world. We asked them to tell us what bothered them most about jet lag. Their responses will sound familiar to anyone who has gone through long-distance travel that crossed several time zones. They included:

- Fatigue or tiring easily
- Trouble concentrating or thinking clearly
- Physical clumsiness
- Decreased daytime alertness
- Trouble with memory
- General feeling of weakness
- Light-headedness, dizziness, or other uncomfortable sensations in the head
- Lethargy or sluggish feeling
- Sleepiness during the day

The basic reason for jet lag is straightforward. Whenever we move quickly eastward or westward, the circadian clock lags behind. In just a few hours, our body finds itself in a new time zone. But the inner clock in our brain is still synchronized to the day/night cycle back where we left from. Let's say you decide to go on vacation to France. Your flight leaves New York's JFK at 6 PM, local time, and you arrive at Paris's Charles de Gaulle the next morning. The clocks in the terminal tell you it is 8 AM. Just the right moment for a morning café au lait and a freshly baked croissant! But your inner clock is still operating on New York time. For it, the time is only 2 AM, about a third of the way into the nighttime sleep cycle. This disparity is so great that the circadian timing system can take days to readjust. In the meantime, you suffer.

Of course the circadian clock does not deserve all the blame for the dislocations that plague travelers. There is the stress of getting ready for the trip, hassles at the airport, anxiety about what's in store at the destination, displacement from your bedroom and its cues for restful sleep, disruptive in-flight noise and air pollution, cramped seating, odd-tasting meals served at odd hours (or increasingly, no meals at all), sudden blasts of bright light in the middle of your attempt to nap. Even so, the disruption of your inner clock is the single most important element in creating jet lag. Fortunately, it is also one that we can learn to deal with.

In the Lab

Chronobiologists think of jet lag as a circadian rhythm disorder. It is in the same category as delayed sleep phase disorder or other problems that stem from conflicts between the inner clock and external time. That means we can study it, and evaluate measures that are meant to correct it, in the controlled conditions of the laboratory. For example, research volunteers sign on for an extended stay in the sleep lab, isolated from the external day/night cycle. During the first week, their sleep period is scheduled for 11 PM to 7 AM. Then, one night, the lights come on at 1 AM. It is as if the volunteer has just taken that flight from New York to Paris and made a sudden move eastward across six time zones.

From that day on, the sleep period is shifted to between 5 PM and 1 AM (the Paris equivalent of 11 PM to 7 AM). Meanwhile, the researchers keep track of pulse rate, blood pressure, melatonin levels, and other physiological measures, as well as alertness, fatigue, mood, and other psychological factors. To study the effects of a westward flight, the procedure is much the same, except that on the day of the simulated "flight," the lights *stay* on until 5 AM, and the dark period then becomes between 5 AM and 1 PM.

It is hard to get to sleep right after a sudden six-hour displacement in *either* direction. The research on melatonin levels shows why. The daily rise in melatonin normally takes place two or three hours before going to sleep. But even after that six-hour shift, melatonin onset stays tied to the earlier light/dark cycle, back before the shift. Easy natural sleep becomes possible only after the melatonin cycle changes to the new light/dark cycle. And that requires resetting the inner clock.

Around the World

What happens to the volunteer in the sleep lab happens also to the traveler. You may be in Paris or Hong Kong, but for the first several days your circadian clock and melatonin cycle are still busy catching up with you from back home. Even without all the other disadvantages—bouts of fatigue, the sense of being in a new place, a delicate stomach, a digestive system that makes urgent demands at unexpected moments—your appetite, energy level, alertness, and general metabolism only gradually get into sync with your new location as the circadian clock resets to the changed cycle of light and dark.

When you get to your destination, the inner clock has no direct way of knowing that it has taken a trip. As a result, it goes on putting out sleep-wake signals based on where you traveled from. At the same time, however, it is being hit by unexpected periods of light and darkness. These cause it to shift around in a jumpy way. It is this combination of factors—a physiological "memory" of where you started your trip and intrusive light when the inner clock expects darkness—that triggers the disruptive symptoms of jet lag.

For the inner clock, the challenge is different, depending on whether we go east or west. On eastward journeys, our rhythms have to adjust to an earlier cycle, while going west they need to drift later to re-sync. Since our circadian cycles tend to run a little longer than twenty-four hours, most people find it quicker and easier to adjust to a westward displacement. On average, we can make an adjustment of about an hour a day on eastward trips, but *two* hours a day on westward trips. This means we can recover from that six-hour Paris-New York flight in three days or so. And owls—those with longer circadian cycles—will recover more quickly going west than larks with shorter cycles. This is because the longer cycles tend to push the inner clock later even without going someplace.

Another factor that makes adjustment harder on eastward trips is the effect of light exposure. That overnight flight from New York gets you to Paris in the early morning (Paris time). But your inner clock thinks it's the middle of the night. Exposure to bright light in the middle of the night shifts your clock *later*, but what you need to do to sync with local time is to shift it six hours *earlier.* As a result, the adjustment will take longer, because your inner clock saw daylight at exactly the wrong time to move earlier.

What to Do?

According to a widely cited rule of thumb, recovering from jet lag takes about one day for each time zone crossed on an eastward trip, and half that long on westward trips. Like most rules of thumb, this one may or may not apply exactly to you. As we said, we would expect early chronotypes (larks) to adjust more quickly than late chronotypes (owls) after eastward flights, while owls have the advantage after westward flights. But even though the one-day/half-day rule is only a rough approximation, think about what it implies. Suppose you live in the U.S., you have a two-week vacation, and you decide to spend it in Europe. That is six time zones east of the eastern United States, nine time zones east of the West Coast. Either way, you will spend about half your vacation recovering from the flight over, and half of your first week back recovering from the return flight. Why would anyone even bother?

There is no secret formula for avoiding jet lag. No mystical crystal or vitamin mixture or herb-filled eye mask will keep you from suffering the consequences of a clash between your new location and your inner clock. However, there are strategies you can follow that greatly *reduce* the effects of jet lag and help you get over it much more quickly. These have been developed in lab simulations and tested in field studies. The details will vary according to such factors as:

- How many time zones will you cross?
- Are you going east or west?
- When does the sun rise and set at your origin and destination?
- How long is the flight?
- Is it primarily a day flight or a night flight?
- What is your chronotype?
- When do you usually go to sleep and wake up?

Complicated, yes, but the underlying goal is straightforward—to bring your circadian rhythm into sync with the day/night cycle of your new location as quickly and comfortably as possible.

Before You Go

The first approach, undertaken before you leave, is to gradually shift your sleeping schedule, and with it your inner clock, toward the time zone of your destination. A simple idea in principle, but there is a drawback. As you get more closely in sync with your destination, you are also moving more and more out of sync with your home. This can cause problems with coworkers and family members. Still, if you know you are subject to severe jet lag, or if you are going to an event where it is vitally important that you arrive in good shape, such as a sports competition, a concert, a business meeting, or a military assignment, the difficulties may be worth it. This approach does work.

Flying East. Suppose you are traveling to Europe from New York. You will have to cross six time zones, so your goal is to shift your inner clock to six hours earlier. If you normally sleep from midnight to 8 AM, that means getting to the point of going to sleep at 6 PM and waking up at 2 AM. So you start a week before your flight. Your objective the first night is to go to sleep at 11 PM and wake up at 7 AM. At 5 PM, six hours before your desired sleep time, you take a microdose of controlled release melatonin. (For an explanation of microdose melatonin, take a look at Chapter 8.) You also dim the room lights and avoid bright light exposure from the TV or computer screen. At 11 PM you turn out the lights and, with a bit of luck, go right to sleep.

When the alarm wakes you at 7 AM the next day, you spend half an hour in bright light, either from the sun or from a light box. That evening, you take the melatonin microdose at 4 PM and sleep from 10 PM to 6 AM. The evening of Day Three you are taking melatonin at 3 PM, falling asleep by 9 PM, and up the next morning at 5 AM . . . and so on. If, as is generally the case, your flight to Europe leaves in the late afternoon or early evening, you'll be able to board your plane, go right to sleep (skipping the meal), and wake up in sync with the morning at your new location.

Flying West. Now you are returning from Europe. Once again you are crossing six time zones, but your goal now is to shift your inner clock to six hours *later*, putting it in sync with the time back home. Because it is easier to reset the clock in this direction, you can wait until three days before your departure to start the procedure. Your sleep pattern has accommodated, so you are sleeping from midnight to 8 AM European time. The first night, you give yourself half an hour of bright light exposure, starting at midnight, and you go to bed at 2 AM. When you wake up at 10 AM, you need to avoid the circadian effects of morning light, so you put on dark sunglasses, or blue-blockers if you need to keep your vision sharp. That night, you use the bright light at 2 AM and go to bed at 4 AM, waking up at noon. The third night you use the light at 4 AM, go to bed at 6 AM, and get up in time to go to the airport, take your flight, and arrive with your biological rhythm already aligned with day and night back in North America.

Of course we realize that these preparatory procedures call for a lot of attention, patience, and determination, not to mention very tolerant partners and family members. And it should be said that while they have been shown to work, they do not work exactly the same way for everybody. Even so, it is worthwhile to get familiar with the principle behind this approach. Even if you are not able to put it into practice completely, if you shift your inner clock only one or two hours toward the time at your destination, you will find yourself that much more awake, alert, and ready to enjoy being in a different place.

The example we've used here is of someone crossing six time zones. You may be wondering what happens with someone who crosses *twelve* time zones, for example by flying from New York to Bangkok. Since that is halfway around the world, time-wise it's the same whether you go east or west. What is the best way to prepare? We can't give a firm answer to that question, but the most effective strategy may depend on your chronotype. Larks may find it easier to shift their inner clock earlier, as if they are going eastward, while owls may find it easier to shift their inner clock later, as if they are going westward. Or, since the westward adjustment is generally easier, maybe everyone crossing twelve time zones should adopt it.

Rachel's Story

Rachel H. was twenty-two and a recent college graduate when she came to our clinic at Columbia. Since her earliest teens she had found it almost impossible to fall asleep at a reasonable hour. During her freshman year at college in California, she often fell asleep at 2 AM and slept until 9:45 AM. By junior year, she had slipped even later, sleeping from 5 AM until after noon. In spite of this, she was not depressed and managed to be a top student. Nighttime was when she came alive, completed her homework, and socialized.

Over the years she had tried various treatments for her delayed sleep pattern, with no success. "I was ready to take on the world," she recalls. "Only I couldn't, because I was asleep all day and awake all night. I didn't know how I was going to be able to work." Now she was planning to do graduate work in London and to make frequent trips from England back to California. Even if she managed to shift her sleep onset from 5 AM to midnight in California, she would have to shift another eight hours earlier once she was in London, then face the problem of major jet lag on each visit home.

To Rachel's astonishment, chronotherapy started helping her right away. "I was incredulous that chronotherapy would do what so many other treatments had promised and failed to deliver on: enable me to sleep normal hours. And then, within six weeks of beginning treatment, I was doing just that. Night after night I took my melatonin, and morning after morning I sat in front of my light box, and day after day I continued to function for what felt like the first time in a decade like a normal human being. My body has the capacity to settle into a normal sleep schedule and this schedule is not as fragile to going haywire as I initially worried it would be."

Rachel's first trip back to California tested her assurance. "My transition back to London time was difficult," she admits. "I didn't resume my light therapy immediately when I returned, since I was on vacation and just trying to relax a bit. A few weeks after returning to London, I began trying to shift my clock back. At first things seemed to be going well. I was waking up naturally a half hour earlier each day and moving my therapy earlier. But then my sleep started to become increasingly erratic. I lost control of the process and was getting very few hours of sleep each night. Ultimately, I ended up seeking guidance from Dr. Terman. Within a week I was back on track and within two I had reached a sleep schedule I was happy with.

"I learned my lesson from that first experience," she continues. "The second time I returned from California, I immediately began readjusting my sleep. Using my past sleep logs and past correspondences with Dr. Terman, I was able to readjust my clock seamlessly, without any setbacks. I now travel between the two time zones several times a year, which means that I have to readjust my body clock again and again. While maintenance treatment in order to keep my body clock from drifting forward is a relatively simple process—melatonin at a certain time each night, an hour in front of the light box at a certain time each morning—adjusting my body clock to a different time zone is challenging and stressful.

"Going east ('forward' in time) is the most difficult. Forcing my body to sleep and wake eight hours earlier is daunting and dreadful. Just like at the beginning of the chronotherapy process, it requires calculating the right time to do therapy and how quickly to shift back. I have to wake up earlier than I want to. I don't get as much sleep as I'd like to. For the week or two leading up to eastward travel, as well as throughout the process of readjustment, I feel stressed about how quickly I will be able to get my sleep on track."

Rachel concludes, "Dr. Terman's instructions for how to adjust to new time zones have been invaluable. I read and reread his advice daily when trying to adjust my clock. As I collect more experiences adjusting to jet lag, I have also found it essential to look back at sleep logs from previous trips to guide my therapy. Despite my experience, overcoming jet lag remains a stressful process, as I have to be diligent about my schedule and constantly vigilant about my sleep patterns, but knowing that I have the tools and the ability to adjust is reassuring. For me, using chronotherapy feels empowering and liberating."

On the Way

You have probably come across advice that claims to help air travelers avoid or reduce jet lag. Some of this advice is common sense:

- Drink lots of fluids during the flight to avoid dehydration.
- Shun caffeinated and alcoholic beverages, even if they are complimentary.
- Eat lightly and only if the timing of the meal fits your personal day/night rhythm.
- If you are going to sleep, use a neck pillow to keep your head from flopping to one side.

We would add to this that if your anti–jet lag preparation calls for you to sleep during the flight, as it is likely to on eastward journeys, take along a good sleep mask and either earplugs or noise-canceling headphones. And test them before you go. Don't rely on getting the earplugs some airlines still pass out free.

Once You Get There

If you are like most people, you probably spend the last few days before a journey simply getting ready to go. There's packing, remembering to stop mail delivery, tracking down tubes of toothpaste that are small enough to satisfy Homeland Security in the U.S., and a long list of other last-minute details. The thought of going through the jet lag prevention routine may be just too much to contemplate. Or maybe you only found out yesterday that you are needed today at a meeting in Brussels or Beijing. For whatever reason, here you are, about to land somewhere several time zones away from your starting point. You would rather not spend the next few days feeling miserable from jet lag. What now?

Once again, the goal is to bring your circadian clock, and with it your sleep/wake cycle, into line with local time as quickly and comfortably as possible. And once again, it matters a great deal whether your flight took you eastward or westward. In either case, you will need to keep in mind when it is day and night back where you came from, because that is what your inner clock is still set to.

Flying East. You need to shift your inner clock earlier, by as many hours as the number of time zones you crossed. Your most important tool for accomplishing this is exposure to bright light during the early morning hours *back home.* For example, you fly from New York to Paris, across six time zones. Ordinarily you sleep from midnight to 8 AM. On your first day in Paris, you want to spend half an hour or so in bright light starting around 7 AM *New York time*, which is 1 PM—early afternoon—in Paris. You also want to *avoid* bright light during the hours that would be the middle of the night back home, because that would shift your clock later rather than earlier. This means you should wear wraparound sunglasses or blue-blockers from the time they turn on the cabin lights until 1 PM Paris time. (For a description of circadian blue-blockers, take a look at Chapter 5.) The next day, you can take off the sunglasses and get bright sunlight at noon Paris time, and by the third day you may not need to do anything further.

You may be able to give your recovery an added boost by using microdose melatonin that closely mimics pineal gland melatonin production, as we describe in Chapter 8. The simplest way to do this is to take the melatonin six hours before your desired bedtime in your new location, then keep the lights dim or wear sunglasses or blue-blockers until you go to bed. If the people around you assume you are someone famous trying to stay incognito, let them think what they like!

Flying West. As we said earlier, adjusting to a westward journey is easier because the internal clock's daily rhythm naturally tends to run a bit longer than twenty-four hours. Here, too, bright light at the right time, and *not* at the wrong time, can be your most important helper. Suppose you fly from New York to Honolulu, crossing six time zones westward. To get the maximum shift in your inner clock, you should get bright light exposure during what is the middle of the night back in New York—or around dinnertime in Honolulu—then stay up until a reasonable bedtime locally. You probably won't need to bother with sunglasses, because you are likely to be asleep during the period when bright light would push your inner clock in the wrong direction. The next day, if necessary, you can search out bright light a couple of hours earlier. But if you're in Honolulu, that may not be a problem—you're probably at the beach!

Is all this a lot to remember? We agree, especially since much of it may seem at odds with common sense. But knowing the reasons jet lag happens allows you to understand the logic behind these steps to counteract it. And complicated or not, if they help you enjoy your journey more, they are well worth it. *Bon voyage!*

17. Chronobiology in the Home and Workplace

What are houses for? During most of human history, the answer to this question has been "to keep out." To keep out wind and rain and snow, broiling heat and biting cold, unfriendly animals and unfriendly humans. If none of these presented a problem, we could simply lay a blanket on a soft bit of turf or string a hammock between two trees. Short of a South Sea island fantasy, though, we need walls and a roof to keep out the elements. And, as a very common but unintended corollary, to keep out light.

If you go to a village in Nova Scotia, or Provence, or the Balkans, or almost anywhere in the world, you can estimate the age of a house quite closely by looking at the size and number of windows. The smaller and fewer the windows, the older the house. Until fairly recently, glass was too expensive for most people, and an unglazed opening would let in rain and wind and intruders. Even some who could afford to put in glass may have thought twice; in the eighteenth century, England, France, and Scotland all imposed an annual tax on windows. The more of them you had, the more you paid. So a solid wall saved money in two ways: You didn't have to pay for the windows, and you didn't have to pay the yearly tax on them.

Our homes and apartments are still designed with a view to taming the elements. In the developed world, central heating and air conditioning are practically standard in new construction. A step or two up the economic scale adds humidifiers, dehumidifiers, air purifiers, and deodorizers—controlled electronically by timers, thermostats, humidistats, and even centralized computer systems. Creating a more perfect indoor living environment is obviously close to our hearts, even as the outdoor environment becomes increasingly foreign and polluted.

Yet light is still largely neglected. Yes, we have more windows than our ancestors, and we have electric fixtures instead of smelly tallow candles and kerosene lamps. But our electric fixtures are mainly designed to conform to whatever is currently considered attractive or interesting. As for the windows, too often they are simply a locale for "window treatments" that serve to keep out the light. Even if they face open countryside and not an adjacent building, the level of light they allow inside the house is only a small fraction of what we would experience if we stepped outside on the gloomiest overcast day. You might not notice the difference, but a light meter would show it clearly.

Our homes may keep us in the dark by day, but by night they keep us too much in the light. Light intrudes from everywhere. It shines through our windows from streetlamps, a neighbor's security floods, the headlights of a passing car, or, as one patient complained, "that damn movie marquee." And inside, we naively impose it on ourselves, with nightlights, illuminated bedside clocks, LED telltales on electronic appliances, and that bathroom light left on all night. This light pollution directly disturbs our sleep quality. It also affects our inner clock during its most sensitive period, the middle of the night. Depending on timing and intensity, it may push your daily cycle earlier, later, or both in succession, with a result the next day that is akin to a minor case of jet lag.

We can control the light we experience, both by day, when most of us need more, and by night, when most of us need less. The technology to do so exists. The fact that we don't put it to use reflects a widespread lack of knowledge. Neither architects nor the general public realize how intensely light and cycles of light and dark affect the working of our circadian clock. Nor do they understand how important the circadian system is in giving us restorative sleep by night and a sense of well-being by day. Poorly managed indoor lighting is a source or magnifier of insomnia, fatigue, poor concentration, and depression. Properly managed lighting can provide a degree of relief to all of these.

In the Home

In today's urban/suburban culture, most of us spend only a trivial portion of the day outside. Even our rare moments of outdoor time do not usually come during the parts of the day when daylight will do the most to synchronize our inner clock to the twenty-four-hour cycle. As a result, lighting inside the home and workplace takes on critical importance. And for increasing numbers of people, the home and the workplace are the same place.

Our starting point for making home lighting a positive source of good health is to take a careful look at the whole house. We need to recognize the ways each room functions at different times of the day and night. In living areas, the goal during the day is to be able to raise the level of light inside the room to what we would experience if we were walking outside on a partly cloudy day. This, we should stress, is quite a lot brighter than you might think. Even an overcast day is about as bright as a TV studio, and as the clouds roll away, the light level becomes ten to twenty times higher.

Enhanced daytime lighting also has an energizing effect, which is direct and almost immediate, as compared to its indirect resetting effect on the inner clock, which is most prominent at the edges of the circadian night. Furthermore, enhanced daytime lighting amplifies the oscillation of the inner clock. During the day, boosts in light exposure reduce the waves of fatigue so many people feel in mid-afternoon or early evening. And, come nighttime—following exposure to enhanced daytime light—the pineal gland produces higher levels of melatonin, which in turn improves quality sleep with fewer interruptions.

Providing an appropriate level of indoor light during the day is not just a matter of turning on lots of bright lights. That would be a recipe for discomfort, headaches, and eyestrain. The light sources must be carefully designed to avoid glare and hot spots. As with therapeutic light boxes, the angle at which the light hits us is important, too. We are more comfortable when the light comes from above and we are not looking directly at it, as if we were out for a walk during the day. One way to achieve this effect would be indirect lighting around the edges of the room.

We also need to be able to control the level of light. During the day, the dosing of intense light is critical. Whatever the light sources are, they must be flexible enough to accommodate the sensitivities of different occupants. The degree of brightness that you think gives you the perfect lift may make someone else in the room feel jittery and agitated. Then as we move toward evening, we need to avoid high levels of light, which could create sleep onset insomnia. That means we need to be able to dim room lights in a controllable way. It also means we should make a point of avoiding late night sessions in front of the TV, computer screen, or tablet.

In the Bedroom

From the standpoint of chronobiology, what happens during the day is important, but what happens at night is crucial—and the space we occupy at night is a bedroom with windows. The transitions from daylight to darkness and back to daylight play an essential role in synchronizing the inner clock and all it influences, in particular the sleep/wake cycle. We are a diurnal species. Our natural propensity is to sleep when it is dark and wake up as the new day begins. If we could achieve a healthy sleep pattern, we would be able to get into bed and fall asleep easily as darkness fell and wake up spontaneously in early light filtering through the window, as the night's last dream episode was winding down. This wake-up would come at a time that was consistent from day to day, with no need for an alarm that goes off without regard for how long we've been sleeping or our stage of sleep at that moment.

One problem with achieving such an ideal pattern in the bedroom is geography. For all but those who live near the equator, the pattern of daytime and nighttime changes markedly as the seasons revolve. In some places, the changes are extreme. People are fascinated to learn that at midsummer, the Arctic region becomes the "land of the midnight sun." Many are also appalled by the midwinter concomitant of days with hardly any sun at all. But even in the middle latitudes of the U.S., nights in the middle of winter are more than five hours longer than in the middle of summer. As for always rising with the sun, that would mean waking up as much as three hours earlier in late June, at the summer solstice, than at the winter solstice in late December.

Modern life does not allow most people to make huge changes to their sleep habits according to the round of the seasons. Even if it did, getting much more sleep in winter and much less in summer would not be kind to our systems. It is simply not practical for us to wake up at sunrise, except at those points in the year when the time of sunrise happens to match the needs of our daily schedule. If the sun is rising too early, we close the shades—but then we may miss the morning light signal altogether. If the sun is rising too late, we are stuck with alarm clocks that deliver an auditory shock regardless of our stage of sleep, and we begin the waking day in dim room light. This does not mean that we have to give up on achieving a healthy sleep pattern. But it does mean that we have to turn for help to new technologies that can replace what the circumstances of our lives have taken away.

The bedroom of the future, in our view, will incorporate computer-regulated lighting technology that can simulate dusk and dawn. At the selected bedtime, the room would fade to darkness and stay dark throughout the sleep period. Toward the end of the sleep period would come a simulated dawn. Gradually the light level in the room would increase, getting closer to the brightness you might experience at sunrise by the time chosen for waking up.

Dusk and dawn simulation has already been studied extensively in the lab and in clinical settings, and the results are often astonishing. Recall the case of Rosie F. (Chapter 7, p. 110). As part of her depression, she was sleeping for long but irregular periods from early evening to late in the morning, then napping during the day. When we treated her condition with dusk-to-dawn simulation, within a week she was sleeping through the night without interruptions, she stopped taking naps, and her depression lifted.

It is time to bring this kind of clinical finding into everyday life. We believe it can be extended to those of us who do not suffer from depression or insomnia but whose sleep and mood fall short of what we would like. In the bedroom of the future, we will have the means to flexibly fine-tune our sleep. We will have overhanging broad-field lighting panels, curved like the sky and controlled by dawn- and dusk-simulation programs that we can adjust to our individual schedules and needs. There are some "dawn alarm clocks" already on the market, but these cannot do the job. They fail to provide the broad field of illumination needed to bathe the bed with light and make up for rolling over away from the lamp. The future bedroom will also need to have effective ways of blacking out the windows to keep out nighttime light pollution and morning light that arrives too early.

In the Workplace

Our experience tells us that employees often find the workplace—whether office or factory—too dim to tolerate, and employers are slow to heed their complaints. Costs and benefits are weighed against each other, and unless there is organized pressure to improve worker conditions, cost saving generally wins out. Inadequate lighting creates a subjective sense of oppressive drudge work and fatigue, and as more people learn about the depressive consequences of short, dim winter days, it can start to feel as if the SAD season continues year-round in the workplace.

The lighting industry has taken some initiatives to design improved workplace lighting. They argue that enhanced artificial lighting on the shop floor will pay for itself in increased productivity, worker alertness and satisfaction, and a reduction in errors, accidents, and sick days. An admirable objective, but so far the results are unclear. Short-term pilot projects have shown positive effects, but that doesn't tell us how long the benefits may last. We know from classic research that some workers react favorably to almost any change in workplace lighting, simply because it is a change. Others may gain real benefit because the enhanced lighting fits their individual needs. Still others may find the change sets off or exacerbates symptoms such as headache, eyestrain, fatigue, and jitteriness.

Our conclusion at this point is that imposing brighter or bluer whole-floor illumination is a mistake. Employees have an enormous range of differences in circadian rhythms and sensitivity to light. As a consequence, enhanced lighting in the workplace will have to be variable across the workday as mood and energy change, and controllable to suit the characteristics of the individual worker. To achieve this, we will need to develop local-area lighting that the worker can control throughout the day while not interfering with those nearby whose choices or needs might differ. Diffuse background illumination from the ceiling would allow for overall visibility without intruding. As office design shifts from classic cubicles to more open spaces, enhanced lighting devices will have to create individual fields of illumination that do not rely on physical partitions. The technology exists already, but applying it in the field remains in the future.

Some employers are concerned about winter blues and depression among their staff. They sometimes ask if they should install light therapy boxes in the workplace. Our answer has been that workplace lighting is important but secondary. A better approach would be to provide their workers light boxes to take home and use each morning *before* they come to work. The inner clock is primed for light exposure at the end of the circadian night. That is the time when the therapeutic effects are maximal. To wait an hour or longer to use the light box at work would provide far less benefit.

Optimizing the Built Environment

Before the techniques of chronotherapy can begin to change our homes, offices, workplaces, and institutions, architects and lighting designers will need to pay more attention to new discoveries in biology. This may sound like a strange marriage of interests and talents, but it is starting to happen. Swiss chronobiologist Anna Wirz-Justice and British-French architect

Colin Fournier recently teamed up to formulate an agenda for this new field, which they published in the prestigious journal *World Health Design*. They don't shy away from acknowledging some inevitable challenges:

- Would a greater awareness of the importance of chronobiology lead to higher levels of artificial lighting illumination and hence higher energy consumption?
- Could this be offset by design principles and guidelines calling for a greater use of natural daylight, and focused, timed, artificial lighting application?
- How can natural biological rhythms—circadian and seasonal—be balanced with a 24/7 society where economic and social requirements take no notice of the geophysical environment?

Yet implicitly, they are optimistic. They suggest that the growing awareness of chronobiology may have as great an impact as the concept of environmental sustainability has begun to have. They also list some of the professions that may feel this impact. Among them are:

- Architecture
- City planning
- Urban design
- Lighting design and engineering
- Transportation engineering
- Landscape design
- Interior design

Those who work in each of these professions will need to develop a basic understanding of chronobiology and work out how to incorporate its insights into their own work.

Wirz-Justice and Fournier have done more than outline the difficulties and challenges. Wirz-Justice has organized a project led by the Center for Environmental Therapeutics to augment the design of two elder care homes under construction in Switzerland with major innovations in chronotherapeutic lighting and negative air ionization, as we have described in this book. The interior group rooms feature a Virtual Sky designed by Oliver Stefani and Matthias Bues of the LightFusionLab at the Fraunhofer Institute in Stuttgart, Germany. This lighting system floods the area with low daytime light levels, including dynamic variations in sky cover color and passing clouds. In the bedrooms, dusk-to-dawn simulators are shaped like a wide skycover and can be programmed by the nursing staff according to each resident's sleep timing and duration preferences, or for therapeutic change in cases of disordered sleep.

The air in the bedrooms, hallways, and group rooms is ionized by newly designed, powerful, silent ceiling fixtures. Patients suffering from depression, dementia, and Parkinson's disease will be able to get intensive treatment with portable bright light therapy panels. When the residents are not using the light boxes, they will be available to the night-shift nurses to help them stay awake and alert. Versions of these new devices—the light box, ionizer, and dusk-to-dawn simulator—will also be made available to consumers for home installation (see *Resources for Follow-up*, p. 301).

Markus Haberstroh's Story

Swiss architect Markus Haberstroh, thirty-five, tells how he became interested in chronobiology, and outlines the promises and challenges it holds for his profession.

"In Switzerland, all young men receive an assignment of community service," he explains. "I was hoping to spend a year abroad on reconstruction aid after the devastating tsunami in 2004, but I didn't get the assignment. So instead I chose a location as close as possible to my home, which happened to be the Basel Centre for Chronobiology in the university's psychiatry department. My assignment was to design their new website, so I began reading their recent publications. It wasn't long before I saw the connection of the two fields.

"It was just luck that I met the right people at the right time," he continues. "They were just beginning to realize that for chronobiology to be helpful in improving everyday life, their lab work needed to be extended to novel architectural design. Before I knew it, I was invited to become a team member of the Center for Environmental Therapeutics, which gave me the unique opportunity to see the interdisciplinary collaboration of their psychologists, ophthalmologists, and lighting designers—with me as architectural specialist. Their feverish enthusiasm, the way they debated new data and fused their specialty interests, and the way they brought me into the discussion was really exciting and sparked my thinking about architectural possibilities that integrate chronobiology's concepts and methods."

In his view, chronobiology is an essential component in architecture. "Up until now architects have not understood or addressed the non-visual effects of light on people—the chronobiological response to lighting, above and beyond the aesthetic qualities. Architects create spaces with simultaneous consideration of different criteria—definition of the space, dimensions, structure, composition, and formal design. The definition of the space is the major step from which everything else derives, and the use of light is one way to define a space. Another factor, which poses constraints, are technical criteria: What is the intended use of the illumination, and how will it fit into energy performance guidelines?"

Much of Markus's recent work has been focusing on chronotherapeutic installations for elder care homes with nursing supervision. "These are exciting projects," he declares. "I am serving as a 'translator' between the specialized planning teams and our chronobiology research group. Planning teams are client-centered, including the user community, architects, electrical engineers, lighting designers, mechanical and construction firms, and the management partners. I am definitely noticing an increase in knowledge, understanding, and appreciation on the user side, especially the nursing staff—the group that will be directly responsible for patient care."

He explains what the project is trying to accomplish. "Patients in hospitals and nursing home residents cannot independently circulate," he points out. "They usually wind up lingering too long in dark areas. The problem is most severe for bedridden patients. We are addressing these problems with optimized daylight exposure, complemented with specially designed bedside lighting devices for strengthening sleep patterns, and simulation of dynamic skylight conditions in the interior group rooms."

The principles that the team is using for institutional settings apply to homes and offices as well. "We are trying to direct more light into the interior from window surfaces, even reaching remote areas with sufficient sunlight. Our use of artificial light is complementary, focusing on the hours before sunrise or after sunset. It is important that lighting installations integrate seamlessly and unobtrusively into the rooms, providing an impression of consistency throughout the day."

As chronobiology becomes more widely known to the public, some problems may emerge. "The lighting industry will of course pick up on the theme and try to market their components under the chronotherapy banner," Markus says. "We already see this happening. The result is a wide range of invalidated products that lack a scientific basis. This is inevitable, and it makes intelligent consumer choices very difficult."

Despite the difficulties of the market, Markus is very positive. "I do not think it will need much persuasion to spread this knowledge to the architectural community," he declares. "I don't see this as innovation. Chronobiological lighting is already well understood from lab, clinical, and field research. Chronobiology is a new topic for architects, but the aim of the two disciplines—to create good and healthy space—is the same."

Gathering the Means

Chronobiology research is a rich source of ongoing innovation. Its insights, methods, and devices can change our homes, workplaces, hospitals, and schools in ways that enhance sleep timing and quality, mood, alertness, and energy—and, indeed, overall health. This is an exciting project, but achieving it will inevitably come with its own difficulties. These innovations will face separate sets of requirements during daytime and nighttime. Still more important, they will have to accommodate individual needs and individual variations in chronotype. One size will never fit all. This last consideration will be the toughest to deal with, because people live, work, study, and sleep together, but the ways they respond vary, sometimes widely.

When we try to imagine built environments that would be chronobiologically sound, some points stand out. They would feature:

- Increased indoor access to natural daylight, the perennial architectural challenge.
- Artificial illumination that safely compensates for any lack of natural light.

- Bright lighting that provides therapeutic effects, linked to the time of day.
- Tapered evening light toward day's end to ease sleep onset (dusk simulation).
- Strictly controlled lighting during sleep, when darkness is the priority.
- Gradually increasing lighting as the sleep period ends, to ease awakening and enhance morning mood and alertness (dawn simulation).
- Non-disruptive lighting for nighttime visibility.
- Filtered lighting to protect from short-wavelength (blue) exposure when it would produce undesired shifts in the inner clock or an inappropriate retinal response after taking a melatonin tablet.

This list may sound daunting, but consider the tools we already have at hand:

- Bright light boxes for close-up, timed therapeutic exposure, whether for its antidepressant, energizing effect or for resetting the inner clock, or both.
- Dynamic whole-room ceiling light systems with spectral variations that simulate outdoor conditions, to boost daytime waking activity and reduce nighttime insomnia.
- Microdose controlled-release melatonin in the afternoon or evening, to synergize with morning light exposure.
- Dusk-to-dawn simulators that overhang the bed, to shape the length and timing of night.
- Blackout shades and curtains, to eliminate light pollution during sleep.

- Filtered glasses, and computer and handheld screen controllers, to block the wavelengths most likely to cause circadian disruption in the evening and at night.
- Amber night-lights to make getting around at night safer without disrupting the circadian clock.

These circadian tools are already available and effective. There is every reason for architects, designers, and private individuals to start putting them to use right now. We can also expect further innovations as designers, engineers, and manufacturers learn about the new discoveries chronotherapy is making in the lab and the clinic.

18. Dawn of a Circadian Science

Our lives are given structure by rhythms from both outside and within. Our world turns on its axis and circles its sun, while the moon circles the Earth. The seasons revolve, the tides ebb and flow. The master clock in our brain, responding to light and dark, proclaims a time to be active and a time to rest, a time to wake and a time to sleep. Each of the trillions of cells that make up our bodies contains minuscule clockwork of its own. We are now moving toward a scientific understanding of how these rhythms work, how they affect us, and how we can use them for our betterment. Applying this new knowledge is the goal and purpose of chronotherapy.

There are many rhythms to life. Some go through their cycle in less than a second, some in weeks or months, and some over the course of years. In this book we have focused particularly on *circadian rhythms*, the biological and psychological processes that are intimately linked to the twenty-four-hour cycle of day and night.

Circadian rhythms touch us even before we are born. The cycles of our mother's hormone production, like her cycles of taking in and digesting food, influence the substances that travel through the placenta and the umbilical cord. As children, we wake, sleep, nap, run around, get giggly or somber or withdrawn, partly in response to signals from our inner clock. At puberty, the rhythms change, but too often the schedules imposed on teens do not keep time with them. If your algebra test is at 9:30 AM, when your brain is still in sleep mode, all that study time last night may be lost in the mist.

As adults, we continue to live in a world with rhythms of its own that often do not work well with our inner clock. We may spend much of the year getting up in the morning and going to school or work in the dark. We may get assigned to the night shift every week, or worse, every other week. We may fly to a distant place and discover that our body still keeps to the rhythms back home. And as we get older, our energy, mood, and sleeping patterns seem to weaken and we drift into dissonance with the cycles of the world.

We are far from alone in our dependence on the rhythms of time. Early in the book we became acquainted with a pale purple flower named *Mimosa pudica,* or "touch-me-not." Every day, as night falls, the leaves of the mimosa fold inward and close up. Then, as dawn draws near, the leaves fold out again—all in response to the plant's inner clock!

Chronotherapy gives us intellectual tools to understand these changes. More concretely, it also gives us a set of powerful tools to deal with them. We can improve our sleep patterns—which has deeper implications than simply getting a better night's sleep. A long list of disorders has been linked to sleep problems, many of them to a delayed internal timing. The list includes attention deficit hyperactivity disorder, obsessive-compulsive disorder, and perhaps Parkinson's disease.

The depressive disorders are the most thoroughly researched so far. We now know that a significant change in sleep—whether oversleeping or not sleeping enough—is an important symptom of clinical depression. Changes in sleep can also serve as a warning sign or act as a trigger for depression. And these depressions, whether they are seasonal, recurrent, chronic, or part of a more complex mood disturbance, such as bipolar disorder, respond to chronotherapy.

Newton's Timely Apple

Sir Isaac Newton is often regarded as the most influential scientist ever. His first ideas about gravitation came to him when he was still in his early twenties, and came—as he told a friend—as he sat under an apple tree and watched an apple fall. Whether the apple fell on his head, as four centuries of schoolchildren have been told, we don't know. We do know that as he thought about the apple's fall, Newton conceived of the idea of a force that would operate on Earth, and on the other planets as well, to govern movement: gravity. To test his hypothesis that gravity existed in space, he asked an astronomer who was studying the movements of Saturn whether Saturn slowed down when it passed Jupiter. It did, replied the astonished colleague, and almost exactly as much as Newton's calculations predicted.

With his theory of universal gravitation and three laws of motion, Newton explained a huge range of phenomena, from falling apples to the movements of the heavens, with as few principles as possible. We call this *parsimony*—the rule that our explanations should not be any more complicated than they need to be to account for the facts. More, he provided the means for others to test and expand the explanatory power of these principles by observation and experimentation. That is what science does. In Newton's case, he went on to investigate the spectrum of colors, develop a law of cooling, and measure the speed of sound, along with setting forth the basic principles of scientific investigation and inventing the calculus. Not an easy act to follow!

We do not mean to say that chronotherapy is about to change the history of science, as Newton did. It is, however, setting out with a similar goal: to explore the ways a variety of facts and observations reveal the operation of underlying general laws. This is a challenging goal. The various problems and disorders we want to fix have different names—depression, bipolar disorder, insomnia, and so on—and different clinical symptoms. We don't know for sure what far-off or recent events, what brain circuits or neurochemicals or genes, underlie them. But it is becoming clear that some of the same basic "building blocks gone wrong" are important in more than one disorder, even though the disorders may appear very different on the surface.

To illustrate this point:

- Imagine a room full of people with colds. They all have the same virus, but some are sneezing and some are coughing. Still others have fever or a sore throat. Some do not have any symptoms at the moment, a few are unlucky enough to have all the symptoms at once. What we see when we look at these people varies widely, but the underlying cause is the same.
- In a room down the hall are people with pneumonia. They all have the same symptoms, but some have bacterial pneumonia and some have viral pneumonia. The symptoms are the same, but the underlying causes are different—and so are the effective treatments.
- Finally, in a very large room—Infection Central—are people with colds, or pneumonia, or both, or neither. They may have symptoms of other sicknesses as well. And they come from different backgrounds and cultures, which sometimes affects the way they experience symptoms and respond to medications.

How are you going to sort things out?

The task is even more challenging with psychological disorders. They don't show up on X-rays, there are generally no lab tests for them, and even the most skilled diagnostician can't detect them by listening to your chest or prodding your abdomen. Yes, very often there are ways to treat the symptoms, and sometimes they work to some extent. But how much better it would be if we understood the underlying cause of the problems and, better still, had the means to treat them at the level of that cause!

Lighting Up Obscurity

Chronotherapy is an evidence-based, experimentally confirmed scientific approach that rests on basic physiological facts. It has shown itself to be safe and effective in treating a variety of problems, from sleep disorders to jet lag to major depression. It has offered new ways of understanding a variety of facts, from the time slippage experienced by Michel Siffre in his cavern to the increased number of heart attacks working adults suffer during the week after the shift to daylight saving time.

So if chronotherapy has so many important applications, why don't more people know about it? People who go to the doctor feeling deeply depressed expect—and usually get—a prescription for antidepressants. Then they wait up to six weeks or more to find out if the pills are working. If they say they have trouble sleeping, they expect—and usually get—a prescription for a sleeping pill. They may also be warned not to use it more than occasionally, even though their sleep problem bedevils them every night. These people do not know there are other ways to treat their problem—ways that are often faster, more effective, and less burdened with side effects. They are not likely to be told. Why? In cases where chronotherapy could work, what is keeping it from being more widely accepted and used?

To understand this state of affairs, we need to look at the different parties concerned. There are patients and doctors, of course, but also hospitals, insurance companies, and those with the most to lose if the status quo changes: the pharmaceutical companies, Big Pharma.

Most patients have never heard of chronotherapy. It is too new a discipline to be talked about in school. There are no commercials for it on TV, unlike the endless commercials for prescription drugs. If patients do find out about light therapy or negative air ionization, it may sound just too weird and unconventional. How can they make the time to sit in one place for half an hour every morning? Their days are already overscheduled. And if they do decide to try it, will their neighbors, friends, and coworkers think they are strange? It takes no time at all to pop a pill, and our culture is one in which pill popping is widely considered the answer to practically any emotional problem.

Not many doctors are familiar with chronotherapy either. Medical schools still cover it only rarely. Physicians do have to take continuing medical education courses to keep their licenses to practice. The requirement is meant to make them stay up to date, but the courses they take are often funded by pharmaceutical companies. We should not be surprised that such courses tend to concentrate on a class of drugs made by the sponsoring company. A program on treatments for depression, for example, will focus on the latest antidepressant medications. There is research showing that cognitive behavior therapy works better than antidepressants in preventing recurrences of depression, but it is likely to be mentioned only in passing at these meetings. Chronotherapy is hardly ever included.

Continuing medical education courses; visits from pharmaceutical reps who have been chosen for their charm and salesmanship; freebies, such as mugs, clocks, and pens—all contribute to an unconscious bias toward medication in general and toward specific drugs. Many doctors agree that these gifts influence their peers, but they also insist that they themselves are unaffected. Not so. When researchers keep track, it is clear that even a small gift or a single continuing medical education course can alter prescribing patterns. Doctors themselves are increasingly concerned. One creative group established an organization called "No Free Lunch!" They have even imprinted ballpoint pens with their "company name." They make a compelling offer: Send us your drug company pens, we'll replace them with ours. No questions asked!

Doctors do face a confusing task. So many medications come onto the market that they can have trouble remembering which ones were approved by the Food and Drug Administration for which uses. In surveys, many do not even know the approved uses for medications they prescribe. They tend to report that the meds they are prescribing have been approved for a particular use, even when they are not. This makes it less surprising that doctors may also fail to follow practice guidelines—those recommendations assembled by the profession to provide the best care. One Harvard study found that many psychiatrists never use practice guidelines. So even if you go to specialists, who are more likely than your regular doctor to prescribe appropriately, there is still a chance that you will not receive the best-established treatment.

These failures are linked to the marketing efforts of pharmaceutical companies. The blockbuster drugs that are still under patent protection (bringing in the highest profits) are often prescribed, whether or not they are appropriate. In one state, over several years more than three-quarters of the prescriptions for a particularly well-advertised, under-patent drug were made without a single published study showing that such a prescription was appropriate. This means that most of the prescriptions for this drug, which has potentially serious side effects, may not have been in the best interest of the patients.

In addition to foot-dragging by physicians and opposition by pharmaceutical companies, patients who think chronotherapy could be the answer to their problems face resistance from hospitals, which may not have the facilities for such treatment, and insurance companies, which do not provide reimbursement.

Dr. Thomas Insel, director of the National Institute of Mental Health, observes that "psychiatric treatments have become largely pharmacological," and adds, "Most worrisome is the relative neglect of effective non-pharmacological interventions."

What Is to Be Done?

It may seem as though chronotherapy faces a difficult future. Difficult, perhaps, but not unpromising. In all walks of life, people resist embracing new ideas ("Let's leave well enough alone")—and doctors are no exception. Chronotherapy is based solidly on careful scientific research, but too often being right is not enough to guarantee acceptance, or even a hearing. We need to turn toward a different model of medical progress, one that puts the patient at the center of the initiative. Call it *Therapy from Below*. Or as Dr. Frederick Goodwin, a former director of the National Institute of Mental Health, put it, "How do you teach psychiatrists? Teach it to their patients."

One of our hopes for this book is to give you, the reader, the concepts and facts to communicate better about chronotherapy with family members and health professionals. You have already taken the initiative by picking up this book. We want you to keep thinking about chronotherapy after you've finished reading about it here. Tag the parts of the book that struck you as particularly interesting and relevant. Explore our supplementary online material (see *Resources for Follow-up*, p. 301). Talk to your friends and family about the inner clock and the surprising effects it has on our lives. This is not just for their sake: When you discuss a topic, you'll clarify it in your own mind and gain confidence about ways to apply chronotherapy effectively.

Dr. Tom Delbanco, a professor of medicine at Harvard, says of today's patients, "They're learning to be more aggressive about their own health care. They're learning to really take us on. They're learning to use us as expert consultants, rather than people who tell them what to do." He knows that patients who really understand their illness and are aggressive in finding out about it—not only from their doctor but also from reading and talking it over with friends and partners—do better in managing their illness.

If you are scheduling a doctor's visit because of sleep or mood problems, prepare in advance and understand the time pressures at each appointment. Make written notes about your symptoms, list your questions (and keep a copy so you can check back later). Take the free, confidential self-assessments (see *Resources for Follow-up*, p. 301) and show your doctor the results. Keep logs of your sleep and mood, if these are posing problems. (General impressions are not enough to go on.) Doing all this will save time during the visit and help make sure you do not forget to bring up something important.

Remember that doctors have feelings, too. Be diplomatic. Say, "I think I wasn't clear" instead of, "You don't understand!" When you bring up chronotherapy, recognize that your doctor may know only a little about the field—you are teaching each other.

Our position is that if you have difficulties related to circadian rhythm disturbances, adjustment of the inner clock should be the first-line intervention, before recourse to drugs. If the clock is at the root of the problem, drugs are unlikely to fix it. Instead, drugs may simply mask the symptoms in ways that fool you into thinking you have found a solution. You will likely have to take the first step by persuading your doctor to give chronotherapy the attention it deserves.

Your inner clock should work for you. It should, and can, help you gain more restful sleep, more energy, and an improved mood. Don't brush aside problems just because you think they are not severe enough to need medical treatment. Can't fall asleep before 1 AM? Not such a big deal . . . unless you need to wake up early or it is making you feel down. Tired all winter long? Not such a big deal . . . unless you are in a high-pressured occupation or it is making you feel down. Do we have to simply accept these as normal, tolerable burdens? Maybe—but they can still be a very big deal for you. Why burden yourself, when chronotherapy may offer simple and effective ways to relieve some of the pressures in your life?

Resources for Follow-up

www.chronotherapy.us

Our field is rapidly advancing. Check this website, created especially for our readers, for the latest updates.

www.cet.org

The nonprofit Center for Environmental Therapeutics offers a wealth of background and guidance for consumers, patients, and doctors interested in light therapy and negative air ionization for sleep and mood disorders:

- Online self-assessment questionnaires for chronotype, seasonality, and depression.
- *Ask the Doctor* forum, where visitors pose questions with archived answers from the center's chronobiology and psychopharmacology experts.
- The center's recommendations for light boxes, dawn simulators, negative air ionizers, blue-blockers, amber night-lights, and microdose melatonin.

- A downloadable form letter that your doctor can use to endorse insurance reimbursement of your light box purchase. Includes an extensive listing of light-treatable disorders, spanning depressive, bipolar, and insomnia subtypes.
- A list of drugs that can cause photosensitization of the retina when using a light box.
- A log form to record sleep, medication, and light therapy times, with ratings of mood and energy.

http://stereopsis.com/flux

Download the computer screen application that protects the eyes from activating stimulation in the evening before sleep.

webapps.nps.org.au/sleepquiz

The University of Pittsburgh's self-assessment questionnaire on sleep quality, presented online by Australia's nonprofit National Prescribing Service.

www.haberstroh-architekten.ch

View Markus Haberstroh's architectural visions for chronobiology.

For Further Reading

Chronotherapeutics for Affective Disorders: A Clinician's Manual for Light and Wake Therapy

Anna Wirz-Justice, Francesco Benedetti, Michael Terman. Basel: S. Karger, 2009

The professional's guide for setting up chronotherapy in hospital and clinical practice. Includes a systematic exposition of basic circadian and sleep science and its translation to clinical trials and application.

"Chronotherapeutics: Light Therapy, Wake Therapy and Melatonin"

Michael Terman and Jiuan Su Terman, a chapter in *The Handbook for the Management of Mood Disorders*, New York: Cambridge University Press, 2012

A detailed guide for dosing and timing adjustments with light therapy.

Internal Time

Till Roenneberg. Cambridge, Mass.: Harvard University Press, 2012

The mechanisms of circadian and sleep timing from a scientist who has worked all levels of analysis—from molecular to mathematical to epidemiological. Contrasts with our emphasis on the bridge to mental health practice.

Rhythms of Life: The Biological Clocks
That Control the Daily Lives of Every Living Thing

Russell G. Foster and Leon Kreitzman. New Haven: Yale University Press, 2005

In a scientist-journalist collaboration, the authors reveal the vital role of biological clocks, the strategies scientists are using to understand them, and the health risks that arise from malfunctioning clockwork.

Seasons of Life: The Biological Rhythms
That Enable Living Things to Thrive and Survive

Russell G. Foster and Leon Kreitzman. New Haven: Yale University Press, 2010

An exposition of the internal calendar in plants and animals, how it is synchronized by passage of the seasons, seasonal adaptations in biology and behavior, and the threat to the system by climate change.

Sleep: A Very Short Introduction

Russell G. Foster and Steven W. Lockley. London: Oxford University Press, 2012

Two scientists from two labs delve into the biological and psychological aspects of sleep, and the causes and consequences of sleep disorders and sleep deprivation.

Winter Blues: Everything You Need
to Know to Beat Seasonal Affective Disorder, 4th edition

Norman E. Rosenthal. New York: Guilford Press, 2012

The now classic but updated explanation of winter depression cycles, with case histories (including the author's own) and advice for management—from light therapy to medications to psychotherapy to diet and exercise, and more.

Index